How the World will Change –
with Global Warming

Earlier books by the author

Reykjavík – Vaxtarbroddur. Thróun höfudborgar 1986
(Planning History of Reykjavík)
Hugmynd að fyrsta heildarskipulagi Íslands 1987
(An Idea on the First Iceland Plan)
Framtídarsýn: Ísland á 21. öld 1991
(A Vision for Iceland in the 21st Century)
Land sem auðlind – Um mótun byggdamynsturs á Sudvesturlandi
í fortíd, nútíd, framtíd 1993
(Land as Resource – On the Development of Settlement Patterns of SW Iceland
in the Past, Present and Future)
Vid aldahvörf – Stada Íslands í breyttum heimi 1995 with Albert Jónsson
(At the Turn of the Century – Iceland's Position in a Changing World)
Ísland hið nýja 1997 with Birgir Jónsson
(Iceland the New)
Borg og náttúra ...ekki andstæður, heldur samverkandi eining 1999
City and Nature – An Integrated Whole *2000* (The book above in English)
Vegakerfid og ferdamálin 2000
(Roads and Tourism)
Skipulag byggdar á Íslandi – Frá landnámi til lídandi stundar 2002
Planning in Iceland – From the Settlement to Present Times 2003 (The book above in English)

On this publication

The author:
 Trausti Valsson *tv@hi.is*
 http://www.hi.is/~tv
Word processing:
 Björk Ingadóttir, Sigrídur Thorarensen and Thorbjörg Sveinbjarnardóttir
Editing and language correction:
 Terry G. Lacy
Lay-out and drafting of drawings:
 Trausti Valsson and Benedikt Ingi Tómasson
Drawings and assistance:
 Karl Jóhann Magnússon, Salvar Thór Sigurdsson and Kjartan Jónsson
Printing and bookbinding:
 Gutenberg ehf

University of Iceland Press
Reykjavík 2006
www.haskolautgafan.is
Copyrights: The author
ISBN: 978-9979-54-727-3

Trausti Valsson

How the World will Change –
with Global Warming

UNIVERSITY OF ICELAND PRESS

Contents

THE FUTURE

Comments on the book

Trausti Valsson is an academic who is not afraid to be controversial. In his new book he argues that global warming can be good for you – if you happen to live in the northern extremities of the globe, which are destined to become the new centre of human civilization. Doubtless, coming just after the UK's Stern report, this conclusion will be fiercely attacked. But it is a useful corrective to much of the current writing on the topic, which fails sufficiently to appreciate that local geographies are going to matter.

Sir Peter Hall
Bartlett Professor of Planning and Regeneration, UCL

Valsson's new book is at the same time provoking and refreshing because it counteracts the prevailing discourse viewing the climate changes foremost as a disaster and instead introduces a historic-geographical perspective in the discussion focusing on possibilities seen from a high North perspective.

Ole Damsgaard
Director of Nordregio
Nordic Centre for Spatial Development

Few issues of international concern have attracted greater attention in years than the likely impacts of global climate change. In his latest book, Professor Valsson offers a new and a daring perspective on the climate change debate, highlighting opportunities that may come about in the Northern regions as a result of global warming. While some of his observations are likely to provoke controversy, Valsson challenges us to think afresh and to get beyond the gloomy predictions and grievance-mongering that frequently characterize ongoing debates over global climate change.

Gunnar Pálsson
Ambassador of Iceland and Chairman,
Senior Arctic Officials of the Arctic Council 2002-2004

It gives me great pleasure to say a few words on Professor Trausti Valsson´s present work. More than 20 years ago I and a couple of like minded collaboraters started to try to get Icelandic authorities interested in Arctic sea routes, in particular the Northern Sea Route. This was based on the conviction that Iceland due to its geographical location could play a role in the foreseeable development towards increased Arctic shipping. The imminent development was due to rapid technical progress in remote sensing, ship-building and information technology as well ever more thorough scientific understanding of processes in northern ocean areas. Our efforts were to a large extent in vain, but a different political scenario in the world since the fall of Soviet as well as the international issue of global warming has, however, opened the eyes of Icelandic authorities.

Professor Valsson is well know in in Iceland for his books, articles and interviews on where future trends will lead the Icelandic society. In this work, his background as an environmental planner has made him qualified to drawing reasonable conclusions from various scientific data relevant to the issues. In his new book Valsson extends his studies to what a warmer global climate may mean for Iceland, and the northern regions of the globe in general. One of the most dramatic consequences is the retreating of the polar ice in recent years.

It should be remarked upon here, that the media frequently publizise theories that suggest that future changes in the Gulf Stream System will lead to the cooling of northern regions, in partiuclar around Iceland. It should be stressed that this is on the one hand a hypthesis "only" and on the other hand pointed at the suggested cooling would only last a limited time in a world of general global warming.
Professor Valsson´s book is a ground breaking work which is likely to be of great use.

Dr. Thor Jakobsson
Head, Marine Meteorology and Sea Ice,
Icelandic Meteorological Office, Reykjavík

In his latest book Professor Valsson, takes a "planners look" at a globe faced with climate change. His main conclusion is that while southern regions may suffer from these changes, the Arctic and Sub-Arctic regions will benefit overall due to increased biological production, influx of southern species and new development opportunities. Hence these regions will become a magnet for human migration and visitation. This is certainly a fresh perspective which will be much discussed and debated in Arctic circles and elsewhere.

Snorri Baldursson
Head of Information Department
Icelandic Institute of Natural History
Member of the ACIA Assessment Integration Team

Introduction

For over one hundred years the climate on Earth has been warming. Actually, cycles of warming and cooling of Earth's climate are not new. Historically humankind has responded to the changes in various ways, though largely by spreading settlements and activities either towards the poles or to higher elevations during periods of warming and, conversely, when the weather worsened, moving away from the poles or to lower ground as the climate chilled and the polar icecaps expanded.

The same migration patterns, towards or away from higher latitudes or higher elevations, have always characterized the biological world in general and not just human migration patterns. Human migrations have been largely a response to the spreading of the climatic and biological zones that humans have adapted to. Today's world of fixed settlements and fixed political borders, however, makes this process of migration more difficult than it has been in the past.

The fixed and settled ways in which societies are organized today, with legally fixed boundaries and clear distinctions concerning land use, make the warming, or the cooling, of Earth's climate more problematic than it was earlier on. Therefore the wish to stop global warming this time around is understandable, especially since, in the opinion of many scientists, the global warming of today is caused by humanoid actions. We are faced with the legacy of the industrialization begun in the nineteenth century with the concomitant result of the extensive burning of fossil fuels as well as the generation of waste products that pollute both surface and air.

One thing seems to be certain: The world is getting warmer. This means a marked change in global weather that will have a very dramatic impact on the natural world and the ways in which we live on Planet Earth.

There are basically three strategies available for human kind to deal with global warming. One is to try to stop, or reduce, the warming by setting quotas on the emission of greenhouse gases (Kyoto). The second strategy is to have both industry and state governments assume the responsibility for creating generation of clean energy and improving energy efficiency (The A6 Nations). There is a third strategy, the one primarily studied in this book and seldom mentioned in public discussion, and that is to prepare ourselves for the migration of people towards cooler areas; towards

Exploration is important. Leif Eiriksson, an Icelander, discovered North America around year 1000.

Arctic areas, grey on the map, are today difficult for habitation but global warming and reduction of the sea ice, will make them approachable and important for future development on Earth.

In 1492 Columbus rediscovered North America, a voyage that started white settlement there.

How a traditional world map is produced:

Step 1: *The two polar regions are cut off.*

Step 2: *The ribbon around the middle of the globe is cut apart in a choosen area.*

Step 3: *The cylindrical map ribbon is now opened until it becomes a flat map.*

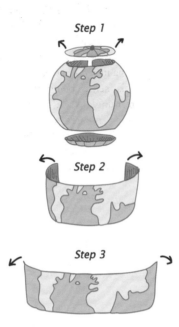

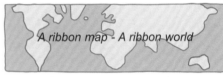

Mercator invented the world-map projection most used today. It gives a distorted picture.

Most maps of Earth only show the globe from the side, where the polar regions are cut off. Here is the process of creating a ribbon map.

Sometimes the two circular maps of the polar regions are added to the flattened ribbon map, that only shows the middle of the globe.

the poles, towards higher elevations, and to cooler coasts.

If the present global warming becomes very extensive – which we will not know until later in the century – the ice and snow of the polar areas will largely disappear during the summer. The retreat of this ice will make the polar areas – which today are almost totally uninhabitable because of ice and cold weather – space that mankind can start to utilize and perhaps eventually migrate to and start to settle – if the heat in the central regions of the globe becomes too excessive.

One of the tasks of this book – which examines the possibility of the creative utilization of the now cold areas – is to introduce the polar regions of Earth to the reader. Today the polar regions are almost unknown to us. This is partially because they have been relatively little explored and partially because the usual maps from which we learn about the world concentrate on the middle section of the world. The commonest map, a Mercator projection, presents the world by cutting off the polar caps (as shown in the picture

above to left), and then the cylindrical ribbon around the central regions of the globe is cut apart and flattened out, whereby it becomes the flat ribbon world map we know today.

In more advanced world atlases circular maps of the two polar regions are added to the ribbon map. The polar maps are, however, so unfamiliar to most of us they look as if they were almost of an alien planet; we do not recognise the shapes and we hardly know any of the terms for the various geographical features.

As seen on the two big polar maps on the right-hand page, most of Earth's landmass is located in the northern hemisphere. If we look at the southern part of the globe; the southern hemisphere in the lower picture, we see that it is almost solely water. In addition to this, what is commonly known among Europeans as "the backside of the globe" is almost wholly water, since the Pacific is so wide that it covers nearly 180° of the Earth's sphere.

The best way of presenting the landmass of Earth in one single image is therefore to show a

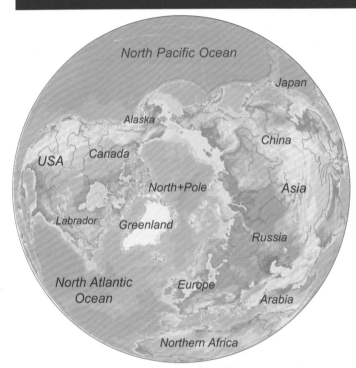

The land areas of the Arctic regions will, to a considerable degree, become habitable with global warming.

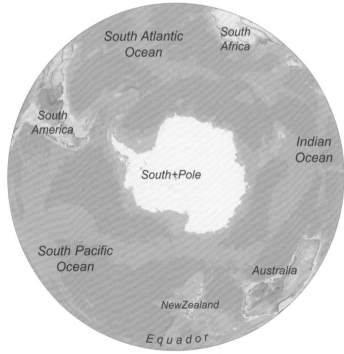

The southern hemisphere is mostly covered by water. Therefore its importance will only grow moderately with global warming.

picture of the globe taken slightly from the north with Europe in the middle. This image shows the Americas in the West and Asia in the East, east and west in this sense being the in-grown terminology we use about directions and areas on the globe.

We easily realize how flawed this old system of orientation actually is from the simple fact that if we place ourselves on the North Pole there is no west or east or south. There is only one direction: south.

Since the main landmass on Earth is in the northern hemisphere and because the areas that will at first start to become habitable with more global warming are areas that have a direct connection to the now uninhabited areas in, for instance, Canada, Scandinavia and Russia, the north sub-polar areas will be of high practical relevance to humankind in the future. In addition, it is predicted that in as soon as thirty or fourty years the Arctic Ocean ice will have retreated so much – particularly during the summer – that major shipping lanes between the North Atlantic and the North Pacific areas will have developed. With most of Earth's landmass in the northern hemisphere, these shipping routes will be much shorter than the shipping routes between continents today through the Suez and Panama canals, not to mention the shipping routes beyond the southern tips of Southern America and southern Africa.

One additional benefit of these newer shipping lanes is that they are not limited by the

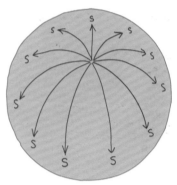

On the North Pole there is only one direction: South, and it belongs to all and none of the time zones.

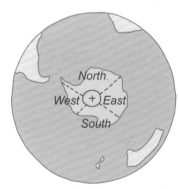

On the South Pole there is only one direction: North. The anomaly: Antartica has a system of directions!

9

A cut through the Pacific puts Europe and Africa in the middle.

A cut through the Atlantic puts Asia in the middle.

Putting N and S America in the middle cuts Asia in half.

All regions want to be in the middle of the world map. They therefore cut the map–ribbon apart in a place that puts them in the middle.

width of canals and therefore a new generation of large containerships and tankers can be put into use in the Arctic routes, making the transportation of goods cheaper.

The southern hemisphere, on the other hand, is of far less interest for the expansion of human activities because of little landmass and the lack of continuity from the other continents to Antarctica. Additionally, Antarctica will neither be habitable nor ice-free as early as the northern regions of the northern hemisphere. This book will therefore mostly disregard possible developments in the southern hemisphere in order to avoid unnecessary complications.

We have now learned how our flawed picture of the realities of our global world is caused by the type of map projections of the world that are most common. The "ribbon world" that appears on today's world maps is nevertheless quite useful for a practical reality of the world of today. Up to now we have been able to safely allow ourselves to cut off the polar regions because they are today mostly unavailable for human use due to the stringent climate: storms, ice and snow.

With global warming, however, the Arctic will become a very active geographical element of the globe, which will mean that our world will no longer be a world on a ribbon extending around the middle but a global world, but rather a semi-global world because the Arctic become added to today's active regions of the world. It will mostly be shipping in the Arctic Ocean that will make the Arctic active, shipping that will activate the the north Scandinavian, Alaskan, Canadian and Siberian regions.

On the page to the right we have four maps looking at the Arctic from four main regions of the globe. It becomes apparent to us, by looking at these maps, first of all, that the Arctic area primarily belongs to very few nations.

The first map shows how the Alaska and Canadian coasts constitute a long stretch of the Arctic Ocean border. On the second map – where the Arctic is viewed from Europe – it is Greenland, Iceland and Scandinavia that are the future frontiers. The third map shows the Arctic area as seen from Russia, which has by far the longest Arctic coastline.

The fourth map shows the Arctic from the North Pacific Ocean. What hits us most is how narrow the gate into the new Arctic world is through the Bering Strait. There are many more gates from the North Atlantic and they are much wider, except for the narrow routes through the Canadian Archipelago. Their narrowness will be problematic for shipping for a long time to come. Practically speaking it is the so-called GIUK gates between Greenland and Iceland, and secondly between Iceland and the UK, that are the entrance from the North Atlantic into the Arctic Ocean area. The ships going through there will mostly, at first, follow the Siberian coast on their way towards the Pacific. After they have sailed through the Arctic Ocean they will enter the North Pacific Ocean through the Bering Strait and will then seek ports on the western coastline of Canada and the USA or ports in China and Japan.

If the warming of the globe becomes extensive these shipping routes will become some of the world's main shipping lanes. This will mean that the nations placed at the described gates: Russia,

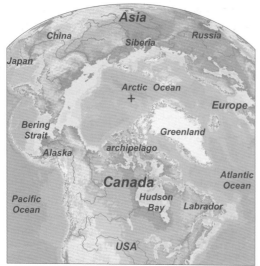

The Arctic seen from four regions of the world.

The Arctic looks very different seen from each of these four regions ...

... Canada, Europe, Russia and the Pacific Ocean.

The Arctic seen from Canada is characterized by the Arctic Ocean and a huge archipelago.

The view of the Arctic from Europe is dominated by Greenland and northern Scandinavia.

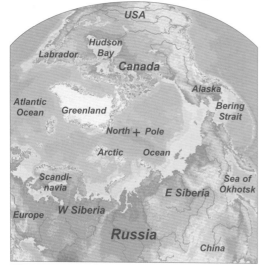

The Arctic seen from Russia is the vast Siberian terrain that stretches over 11 time zones.

The narrow Bering Strait is the only entrance from the Pacific space into the Arctic Ocean.

the USA (because of Alaska) at the Bering Strait and, to a lesser degree Canada, Greenland, Iceland and Norway, will acquire a strong geopolitical position at the Atlantic gates. As the routes at first will follow the coasts of Russia, Canada and Alaska over long stretches, their geopolitical position will be very powerful.

In addition to the possible control of these very important shipping routes in times of crisis, the depletion of the resources of today´s world will make the natural resources of the Arctic regions – be it oil, gas, timber, fish or minerals – very important for the future, putting the Arctic coastal countries in a very strong position. In times of geopolitical conflicts the nations bordering the Arctic areas, and bordering these narrow gates of world shipping, will have a very strong position, somewhat similar to the position of Gibraltar, Panama and the Suez areas of today.

It may come as a surprise to the reader that this brief sketch of the main task of this book does not pay much attention to the many terrible con-

Let's Embrace Change!

A New Way of Life

Modern man
Is static in his ways.
This is in conflict
With the nature of the world.

Time changes everything,
But Time is not accepted.
And we don't accept the changes,
That time brings.

The world is changing,
Fast and profoundly.
Today, global warming
Is the catalyst of change.

Not to fight change
Is the lesson to be learned.
The new way of life
Is to embrace change!

sequences of global warming for nations that are centrally located on Earth. The reason is that this book rather focuses on the new possibilities that mankind is offered by the warming of the Polar regions. This is not because the author is comfortable with what is happening with global warming but rather that this is a viewpoint that has been very little studied in the literature, a study that can even can be termed a "study of the positive impacts of global warming". Space does not allow an extensive review of the many negative impacts of global warming, nor is it necessary as the negative impacts have already been described extensively in numerous books and reports, whereas the positive impacts have hardly been examined.

It should also be understood that the author is not a climatologist but a planner, so he has no possibility of making a judgement as to how fast the globe is going to get warm, or how warm it is going to get. The author's approach begins with the assumption – provided in many scientific reports – that it is going to get warmer. For the sake of clarity he extends that projection far into the future. In the future it is possible – if warming continues for a long time – that many central regions of the globe will actually become uninhabitable, so the only option left to humans is – as in earlier times of excessive warming on Earth – to migrate towards the poles.

The author wants to stress that he thinks that it is sensible to try to reduce the speed of global warming by reducing greenhouse gas emissions, for example by complying with the Kyoto Protocol, and he also, of course, thinks that every attempt possible should be made to create clean energy, energy that produces little in the way of greenhouse gases. If that happens – which at the moment does not seem very likely – the problem of extensive global warming might largely be solved by pollution-free energy resources.

However, in the case that neither the curbing of greenhouse gases nor enough clean energy is developed, it may well happen that migration to the Arctic regions (and to higher ground), will be an option and a strategy that needs to be studied thoroughly as a response to the problems brought upon the world population by global warming.

Very often in history we have experienced human migration, often as a response to changed climatic conditions. An example is the migration of Northern Europeans to America in the late nineteenth century as a response to a very cool period.

This time around we will not have a largely unsettled continent (the Americas) at our disposal, at similar latitudes, but rather the new areas that will become habitable by the warmer climate are to the north (and south) of the inhabited areas of the globe, primarily in Northern Canada, Greenland and Northern Scandinavia and Northern Russia. A natural migration would then be, for instance, that Europeans would move up to the newer areas of Greenland, Iceland and Scandinavia and people living in the south, in Northern Africa and the Middle East, would move north into southern Europe and southern Russia, as has happened so many times before in the history of the world.

The political reality of today's cultural and political borders will make this seem to be a very problematic migration, since the way we think about our place in the world as been fixed to a certain physical location, and we also think it almost impossible that these unlike cultures are able to integrate in a peaceful way. This calls into question an examination of a potential future perspective.

This type of migration, however, might be the only solution mankind has and we, possibly, will have to change our mindsets to fit this new reality. We might have to abandon our love of the static view of the world and we will possibly have to look at these changes as possibilities, even solutions, and not problems.

We also might have to abandon our current idea that different races, different religions and different cultures should be kept apart in separate regions. And we might need to develop political and cultural solutions that are far more tolerant of any remaining racial or religious differences.

1 Warming: The Catalyst of Change

1 How Global Climate Changes

It is no longer a prediction, it is a fact: the globe is getting warmer. Meteorologists, climatologists and other scientists have already described the changes in numerous articles and reports. This book will take that knowledge and these projections further and describe which areas of the globe will become more liveable, and which areas will become less liveable.

Here we already have one of the main points of the book: there is *a balance!* Global warming entails both positive and negative factors – whereas most reports indicate global warming as being solely negative. Again: there are areas – towards the poles – that will gain hugely from global warming, as there are also areas that will suffer enormously from the consequences from global warming, mostly because today they are densely populated.

The changes that come with global warming will be far reaching. It must be kept in mind, however, that climatic fluctuations on Earth are not new. Regions close to climatic limits – for example in terms of cold or dryness – are of course much more vulnerable to climatic fluctuations than the regions situated in a more benign climate. One example, though not widely known, is that several times a drop in only one to two degrees Celsius had a catastrophic outcome for Iceland in its eleven hundred year history. In contrast, Icelanders have regarded warming by one to two degrees as a gift from heaven and the country prospered accordingly.

Occasionally in the history of Earth climatic warming or cooling has been much more extreme, even more extreme than is projected now. Some of the oldest cultures of Earth, like China and Egypt, have a recorded history of what has happened with excessive warming, or cooling. Looking much further back in time, deep into the Earth's records in fossils and rocks, we see that there have been periods of excessive warming where all polar ice disappeared and redwood trees grew in areas that are now Arctic. On the other hand, there have been very cold periods, during the Ice Ages several thousand years ago, when the Nordic ice sheet reached down to Berlin and the ice in North America stretched as far south as present-day Kansas.

Given the adaptability of living organisms they have been able to adapt or evolve to cope with these changes. Humans, too, when it became their time on Earth, have made significant adjustments

The climate of the globe is getting warmer, because emissions accumulating in the atmosphere trap the warmth of the sun's rays, as happens in greenhouses.

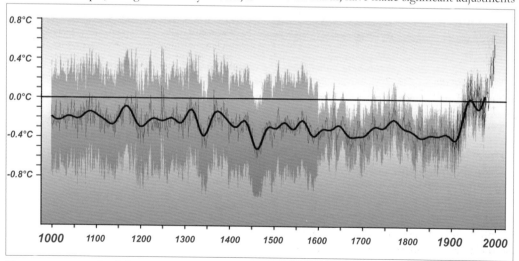

Temperatures on Earth have been rising sharply in the last 100 years. The straight line shows average temperatures in 1961–1990. This average is far higher than the average of the past 1000 years.

The retreat of the Arctic sea ice in the past three decades is a clear indication of global warming.

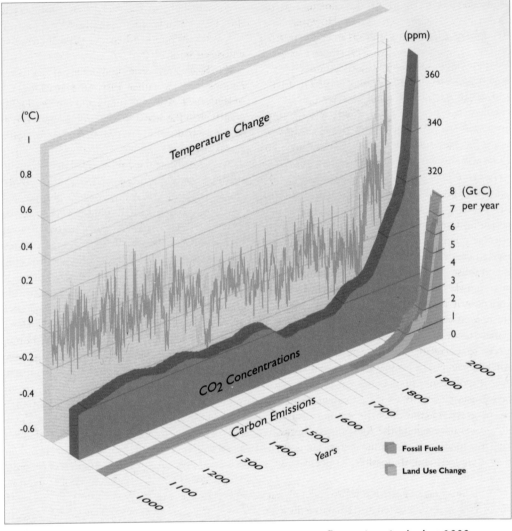

The pink zig zag lines on the blue panel show the temperature fluctuations in the last 1000 years. The brown panel of CO_2 concentrations shows a clear correlation between the two. The smallest panel in the front shows that the increase in CO_2 is caused by andropogenic carbon emissions.

to the advances and retreats of the polar ice sheets. Polar bears even survived the warm periods, and humans in the northern regions spread south with the advances of the ice and north again as the ice retreated.

The cause for these extreme climatic changes is not of central interest for this book; it is assumed that nature is dynamic and constantly changing. Instead the book presents likely changes in terms of human settlement and activities that will result from a continuation of the present global warm-

ing, intended as an outline of factors and potentials that should lead to further discussion.

This first chapter reviews the main theories regarding why the globe is getting warmer at this time. No single natural occurrence seems to be a prime factor in causing the warming. Most scientists therefore think the warming is caused by greenhouse gases that have accumulated in the atmosphere from the start of the industrial revolution and, in fact, there is a close correlation between the increased amount of greenhouse gases and the warming of the climate – as the graph to the facing page shows. However, there are some noticeable exceptions in some areas of the world, like the cooling of Northern Europe in the late nineteenth century until about 1920, and then again some cooling from ca. 1950 to 1970.

Sceptics point out that a correlation is not proof – and that the climate has previously often heated up in a short period of time, also before the industrial revolution when there was no major andropogenic release of greenhouse gases.

Again the point should be stressed that the cause of today's warming is not of central interest for this book – but rather the impact the warming is going to have on human existence. On the other hand, if we think we can be certain that there is a very close correlation between greenhouse gas emissions and warming trends, that knowledge would be useful in many ways. In this case it is rather easy to create scenarios of likely increases of greenhouse gases and then to extend the warming trends in a way that they continue to correlate closely with the amount of greenhouse gases.

The political consequences of linking greenhouse gas emissions to the warming trends are rather grave. Therefore a large group of industrial nations have signed the *Kyoto Protocol* declaring that it is their goal to stop the increase in undesirable emissions and, if possible, to decrease such emissions.

In the first stage of the Kyoto Protocol the target of each nation, or union, had been to reduce the amount of emissions by the year 2002. Only very few Kyoto members reached that goal; Germany is scoring best with minus 18.6%, mostly because of the unification of Eastern and Western Germany which led to closing many of the "dirty" industries in the east. This reduction led to reduced emissions for the unified German state.

Many of the Kyoto nations have had very little problem keeping emissions at bay because they are growing out of the industrial era into an information and high-tech society.

The European Union has one common Kyoto quota which allows a reduction that is reached in one EU country to be used to allow an increase in the quota in another EU country which is still in

its industrial phase. This we can see in the table to the right, where for example, Portugal was allowed a 27% and Greece a 25% increase in emissions until 2012.

We have now seen why the old industrial nations have a rather easy time reaching the objectives of Kyoto, in contrast to the developing countries in the world. It will probably be seen as a rather unethical request of the industrial countries – that built their wealth without having to pay anything for their emissions – to try to pressure the developing countries into signing the Kyoto Protocol after 2012.

It should also be kept in mind that most Kyoto countries are countries of limited energy resources. This means that it is of huge national interest to them to reduce their dependence on carbon fuels and, instead, to try to develop renewable energy sources, such as solar and wind power.

Another reason why the densely populated industrial countries are more willing to put limits on industrial exhaust, is that pollution is more localized in the densely populated industrial countries than in most developing countries. Still another reason why the industrialized countries are willing to impose emission quotas on themselves and their industries, is long-term thinking that recognizes that energy prices will rise in the near future because of lower supply in relation to increased demand. By imposing quotas, the governments are pressing the industries to install more energy efficient technologies. This fuel efficiency will result in a competitive advantage on the world market as energy prices continue to soar.

In addition to the developing nations there are industrial nations that have not been eager to sign the Kyoto Protocol. Russia, at first, did't want ot join but did eventually, probalbly because – like other heavy industry nations in Europe – it can envision that it can reduce heavy industry and thereby gain valuable emission quotas.

The USA has not been willing to sign the Kyoto Protocol, probably for political reasons because the USA is losing industries at a high rate to developing nations. Additionally, it is harder for the USA than for many other countries to bear the burden of these quotas – except for a

Country	Kyoto	2002
Belgium	-7.5%	+2.9%
Canada	-6%	+20.1%
Denmark	-21%	-0.4%
Germany	-21%	-18.6%
Greece	+25%	+26%
Iceland	+10%	-4.2%
Luxembourg	-28%	-19.5%
Portugal	+27%	+40.5%
Russia	0%	-39%
UK	-12.5%	-14.5%

Nationally accepted Kyoto goals for release of greenhouse gas emissions 1990–2002 in % ...

...and the actual addition (+) or decrease (–) in release of emissions in1990–2002... also in %.

Most Kyoto countries are growing out of heavy industry, giving them opportunity to reduce emissions.

long adjustment period – since US industry and their way of living are based on and adjusted to very low energy prices. The car-based planning of US cities makes it almost unthinkable to commute without cheap gas; most are too sprawling to support public transportation.

The eleventh conference on the Kyoto Protocol was held in the autumn of 2005 in *Montreal, Canada,* where a new phase of the protocol was adopted by the Kyoto nations. It was very important that Russia had joined; otherwise the protocol would not have been ratified. Despite the inclusion of Russia, the Kyoto nations only account for one-fourth of the greenhouse gas emissions of the world.

As for possible world reductions in emissions, the greatest importance concerns what will happen in the USA, China and India. In January 2006 these nations, together with Australia, South Korea and Japan, held a conference in *Sydney, Australia.* There a ministerial conference of these six nations – that already today, account for half of world emissions – established an organization parallel to the Kyoto group, called the *Asia-Pacific Partnership on Clean Development and Climate.*

The main characteristic of this partnership is that it is not going to agree upon any binding goals or quotas, but rather it is a partnership based on a common policy and research in eight main areas. Many environmentalists maintain that this partnership was formed mostly in order to have an excuse to not take part in the Kyoto framework. However, it has to be admitted that the plan presented in Sydney is rather impressive and surprising. These nations proclaim that they will be able to continue and even increase their economic growth and energy use and at the same time, reduce their emissions significantly. Partly this should not come as a surprise because their crude industries can rather easily adapt environmentally sound and efficient technologies as such changes have become cost efficient with higher fossil fuel prices.

In the Sydney Conference these A6 nations presented objectives that seem to hold some promise in terms of gaining significant results in eight main areas that have a great deal of influence on the quantity of greenhouse gases emitted into the air. *These eight areas are:* 1. cleaner fossil energy; 2. renewable energy and distribution generation; 3. power generation and transmission; 4. steel; 5. aluminium; 6. cement; 7. coal; and 8. constructions and appliance in general.

The main difference between the A6 and Kyoto approaches is that Kyoto defines limits and distributes quotas, whereas the A6 partnership does not put limitations on growth or binding quotas on emissions. A certain difference in philosophy is manifested in the two approaches: The Kyoto nations, their leaders being primarily European, opt for the cult of regulations, whereas the Asia-Pacific nations put their faith in the belief that the industries themselves will have the urge to curb emissions and reduce the use of energy. This, however, is probably not going to happen to any significant degree unless fossil fuel prices rise even more than is the case today.

It is a characteristic of the Asia–Pacific group that their agreement expresses a strong belief in the ability of free industry to create new technologies, and in the run up of the Sydney meeting *Ian Macfarlane,* the Australian Minister of Industry, said "the reality is that new technology will deliver three times the savings in greenhouse gases as the Kyoto Protocol will."

The establishment of this new group of large industrial nations that constitute about half of all emissions in the world today does probably mean that they are not going to enter the Kyoto Protocol, neither in 2012 nor in the further future. That will probably mean that the Kyoto nations will not be willing to ratify an extended agreement about quotas and emissions after 2012, since such an international agreement needs to have more than one fourth of the emissions in the world within its frame, so it would make sense for them to put restrictions on their industries and economies.

It is probably also true that by 2012 the Kyoto nations will already have gained the most they can out of the quota pressures they have put on their own industries in terms of increasing fuel efficiency to be more competitive in the future and that they will probably, at the same time, have secured such a degree of air cleanness that the microclimate in cities will be acceptable in most places.

The 11th conference on Kyoto was held in Montreal in the autumn of 2005.

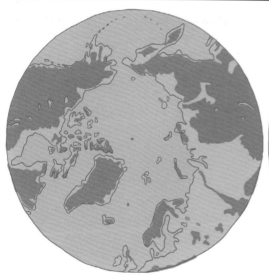

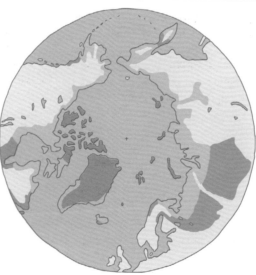

Northern areas above 200m (brown) are not fit for habitation. The low areas fit differently well for habitation, see the picture to the right.

Areas that will become inhabitable are light green, and areas that will become very suitable for habitation dark green. Grey shows uninhabitable areas.

As stated earlier in this chapter, the assessment of what the nations involved in this debate are in fact thinking, also needs to consider *what the geopolitical consequences will be:* what areas and nations would gain and what nations would lose geopolitically with global warming.

Looking at Russia first – with its wide, cold, northern terrains – we at once recognize that it does not to need to be very negative for Russia that it is going to become somewhat warmer there; the Russian population can rather easily, over a long period of time, spread to the more northern areas. This also holds true for Canada.

In general, the old industrial nations are mostly northern nations of rather cold climates, so one is tempted to doubt that they are, in fact, so very worried about the climate getting somewhat warmer, since warmer temperatures will mean a more comfortable outdoor life for them. In assessing the impacts of global warming one must, however, also keep in mind that it is not only increased warmth that comes with global weather changes; it is all kinds of alterations in weather patterns such as changes in precipitation, where some areas may experience less and others more precipitation. More precipitation in some areas, for instance means heavier snow loads in the winter. In Europe recently this has already meant roofs collapsing from the unexpected weight of snow and increased rainfall that has meant greatly increased flooding.

Another consequence of global warming and the melting of glaciers and the expansion of the volume of the oceans, which leads to the prediction that the world's sea level will rise by about 50 to 90 cm in this century. To combat the resulting coastal flooding will be very costly – and here again the industrial nations, the rich nations, will have more funds to deal with the problem, whereas the poor countries have little extra money for the necessary coastal defences. In addition to this, some of the poor counties that have settlements in very flat estuaries, like Bangladesh, are so densely populated that they lack space to move people out of the very dangerous flat terrains. It needs, however, to be mentioned that today's flooding there is caused by a more rapid runoff in the rivers because of deforestation and hard surfaces.

Various other types of changes that will happen *in the natural environment* with changing climate patterns are discussed in the next section. The section after that deals with the impact of global weather changes *on human habitats and activities.* The fourth section of this chapter on the primary impacts of global warming describes in more detail *three options that are available for dealing with global warming.* The first option, or strategy, is the

Sea level may rise 50–90cm in this century. Most erosion occurs where ice retreats from coasts.

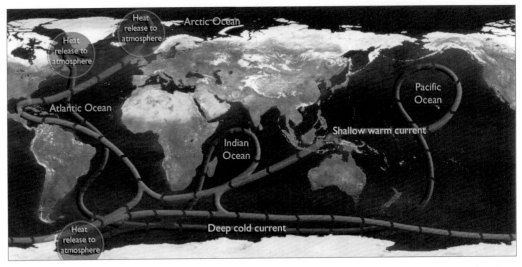

This picture shows the main circulation patterns of the world's oceans. The warm currents absorb heat in the warm areas of the globe and release it to the atmosphere in the cold areas. There are theories say that this conveyor belt may slow down or otherwise be disturbed.

strategy of the Kyoto group. The second strategy is the Asia-Pacific way of approaching things. The third strategy is *the natural way,* that is, *to accept, and go with, the changes* and *move people and settlements away from areas that are becoming too problematic* to the areas that are going to be more benign and better fitted for habitation with global warming. In terms of looming higher temperatures, better areas will lie further north and south on the globe, in higher elevations and along coasts.

Finally, it should be mentioned that even though scientists in general agree that the climate is most likely going to become very much warmer in the future there are theories dealing with other important factors and changes. One deals with the *conveyor belt of ocean currents,* postulating that disturbances in this system of ocean currents that transports warmth to the colder areas of the globe might change the climatic future in individual localities or regions of the world – for instance in Western and Northern Europe by the displacement or slowing down of the present Gulf Stream. But these are just theories and data to verify them, or refute, are largely missing. (See discussion on page 127).

Some scientists have also argued that the cold ocean currents coming from the Arctic Ocean and passing along the east coast of Greenland will become so much lowered in salt content by the melting ice that the currents will no longer be heavy enough to sink to the bottom as fast at southern Greenland as they do today. This would mean that these cold currents would flow on the ocean surface further south, along the eastern coast of Northern America, and would, logically, lead to a cooling of that area. This scenario was visualized in the film "The Day After".

The possible changes in currents are hard to monitor and predict. Some monitoring stations have been placed to measure the salt content and the speed of these currents, but studies like these have to be made over decades in order for them to have any statistical relevance, since short term changes cannot reliably tell us much about changes over longer periods.

The main predicament, in spite of regional differences, will in general remain: *The world is going to get warmer.* It is of considerable significance that the warming will become much greater towards the poles – especially in the Arctic region. This warming will make the Arctic more ice free and more accessible, whereas, at the same time, many of the central regions of the world will be prey to the negative side of this development, i.e., a climate that is too dry and warm, with expanding deserts and also, at the same time, flooding of many highly populated areas because of the rise in sea level.

The film the "Day After" visualizes the theory that a cold current will mean the cooling of NY.

2 Changes in Natural Environments

We have now reviewed in general terms how the global climate is expected to change with global warming. The next logical step is to ask how these climatic changes will *change natural environments on Earth.* In order to be able to give an answer to that question, we now need a more detailed picture of how the various patterns and parameters of the climate will change.

As we obtain a clearer picture of how climatic patterns change, we can draft a picture of the changes in the natural environments that follow, meaning both the physical environment, like landscape features, as well as the vegetative cover.

As we know, climate and vegetation form the framework determining what types of organisms can survive and thrive in a specific environment. The third step in *this logical interdependence* is to ascertain how humans adjust to the climatic and environmental features, because if climatic and environmental features change, the lives of humans, at least to some degree, must also change accordingly. *Such changes in human habitats* are the subject of the next section of this chapter.

The study of global warming in the literature, historically, started with a study of climatic changes. This has been continued with studies on how these climatic changes will impact natural environments. The third aspect, *the impact on human settlement structures and habitats,* has been much less studied. This subject area is the main theme of this book. Within this frame of human habitation, the book mostly deals with changes in human habitats on a global scale, focusing mainly on present and upcoming changes in the northern regions of the world.

Many scientific studies have been done on climatic change. The most prominent ones are the *IPCC studies of the Intergovernmental Panel of Climate Change* that works under the aegis of the United Nations. Its first report was issued in 1990 and the third in 2001. The next report is due in 2007. These reports look at the changes from a global prospective, whereas individual countries, like the USA, have published more detailed reports on the impact of global warming on their area.

The first detailed report on global warming for a large section of the globe was written by the *Arctic Council,* which is an association of the eight countries that boarder the Arctic area. The work on this report took five years and involved about three hundred scientists. Each nation shoulders the chairmanship for two years, and at the end of the Icelandic term in 2004 a summary report from this work was published at a ministerial conference in Reykjavík.

The scientific report, which is about 1000 pages, was published in 2005 by the Cambridge University Press under the title of *Arctic Climate Impact Assessment,* or ACIA. The general scientific methodology of *Environmental Impact Assessment* (EIA), applied in this report, was largely developed in the last few decades. This methodology at first was mostly used to assess large individual projects deemed to be likely to have an impact on the environment. In EIA an assessment is carried out and a report written in order to understand and describe what impacts the project is likely to have on the environment. In the ACIA report this assessment methodology was used for describing the impact of changed climatic parameters and patterns on various environmental, human and construction features *in the Arctic.*

This description is of especial interest to the whole world because the Arctic is warming much faster than areas at lower latitudes. *The reasons for this speedy warming are:* 1) Reduced snow and ice; 2) More of the energy trapped under the Arctic skies goes into warming rather than into evaporation; 3) The atmospheric layer of the Arctic is shallower, so it will warm more; 4) As the sea ice retreats more, more heat will be absorbed by the oceans; and 5) Alternations in atmospheric and oceanic circulations can increase warming, and as the sea ice retreats more, the warm currents coming from the south will have easier access to northern areas.

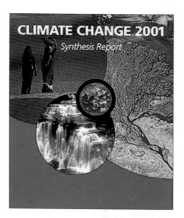

The 3rd report of the International Panel on Climate Change (IPCC) was published in 2001.

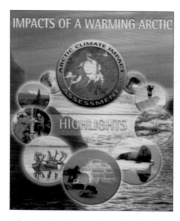

The Arctic Council published the Arctic Climate Impact Assessment (ACIA) in 2004.

For technical reasons, reports like the ACIA report have a projection range to the year 2100. At the end of this period it is projected that the temperature will probably have risen by some 4 to 7°C over land areas in the Arctic. Northern areas, like Scandinavia, will in the same period, only have experienced a 2°C increase and areas in the middle of the globe probably only about 1 degree, which, however, are high rates for countries that are already close to the limits of what many organisms can tolerate in terms of heat.

With the result of so much additional energy trapped under the atmospheric layers, with global warming all climatic processes on Earth will accelerate. There will therefore, in general, be more precipitation and more strong winds. Another feature that we have already observed, not least in the last 15 to 20 years, is that various climatic patterns on Earth have been changing. This means that in geographic terms – even though in general there is going to be more rain – some areas may experience less precipitation as rain or snow. Another feature – a feature that we have already observed, not least in the last 15-20 years – is that climatic extremes are increasing. There are already periods of high precipitation and high winds in many areas and, conversely, periods of droughts and calm in other areas. The third main change in climatic patterns, in addition to precipitation and wind, is the change in the climatic cycle of the year. In general, winter, especially late winter, will become warmer and wetter. Ecosystems and organisms need to adjust to environmental conditions for their well-being. Plants and organisms need to be adjusted to environmental conditions in order to survive. If conditions change, ecosystems and organisms suffer. During long spells of drought, as an example, certain species of plants can die and, conversely, in long spells of heavy rainfall, certain arid species will suffer.

Of course, the natural environment frequently involves cycles, seasonal and others, so that specific organisms have a tolerance for specific climatic alternations, and if the changes do not happen too swiftly, *ecosystems can adapt to new environmental conditions* – though they may emerge as changed ecosystems. What is already now happening with the warming of the globe is that a large number of organisms are migrating towards cooler areas that they are better adapted to flourish in, i.e. there is a shift in vegetation and climatic zones, and both the plants and animals that are associated with these zones are shifting to where climatic factors are optimum for these organisms.

As the world climate, becomes warmer and wetter, especially in the high North, vegetation there will increase and the now arid and sparsely vegetated Arctic tundra will slowly start to become like the northern vegetation zones of today. Boreal forests have already started to extend into the Arctic tundra and the grassy vegetation of tundra into the polar deserts. The time frame for these shifts will vary throughout the Arctic. In places where rich soil exists, changes are happening faster than in other places.

The lifestyles of the peoples of the Arctic will be dramatically changed by global warming.

The figure shows seven projections for global warming from 1990 to 2100. The average for the globe will be about 3°C in 2100.

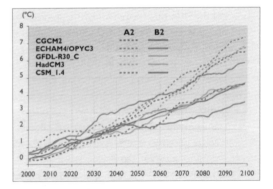

Arctic temperature will increase from 2000 to 2100 from 4 to 7°C in 2100. The trend points to still more warming after 2100.

Present on-going measurements and projections based on them, plus understanding of the range of climatic factors required by specific organisms, make it possible to map how the Arctic will probably look in terms of vegetation in the year 2100. This northward shift in vegetation zones – along with rising sea levels and thawing of therma frost at coasts, means extensive coastal erosion and the flooding of lowland areas. As a result of what has been described the Arctic tundra is projected to reach its minimum extent for at least the past 21,000 years.

These extensive changes in the Arctic will have a huge impact on the ecosystems and the animals that now live on the ice or on the tundra. This will mean a reduction in the number of polar bears, ice-dependent seals and caribou/reindeer. This is certainly nothing that we can be happy about but this, in fact, is how things have always happened with global warming, and as we approach the North Pole area, there will be no farther cooler North for the northernmost species to migrate to. In addition, the reduced size of the habitats of these species, will be invaded by other species. It is of no use to bemoan these changes. Though these changes will be monumental, even catastrophic for some species, as well as for homo sapiens, *such changes have happened many times before in the history of the world.* Just as some species will suffer, others will thrive, and in general the whole Arctic region will become more productive, both the land areas and the ocean.

Highly valuable fish stocks, like cod and herring, will migrate northwards because less sea ice cover and the resulting more sunlight will increase the production of plankton, making these – now prohibited – areas of the Arctic, a premium fishing ground.

The effect of the disappearance of the permafrost in the Arctic is already visible in many areas. With melting, soils that have been firm become softer, which means, among other things, that the roots of trees lose some of their hold in the ground and start to lean to the side. This has started to happen and in some places today has produced what is called a drunken forest, since trees look like drunk people, leaning in different directions.

The reduction of permafrost also means less carrying capacity of the soil. As in many areas of the Arctic, roads and buildings can, with special construction techniques, be engineered on such frozen soils. In some cases, such as the Richardson Highway in Alaska, the roads can be used in winter only when the land remains firm under the weight of the vehicular traffic. With thawing, foundations start to subside at different rates and construction can tilt and be damaged.

On the positive side, with global warming rivers become more ice-free so that larger boats will increasingly be able to navigate the rivers as well as the coastal waters. The increased industrial activity that comes with easier transportation will mean large changes in the environment. Industrial pollution and oil spills will become more frequent and the impact on the fragile ecosystems of these cold climates will be graver than in many more southern areas.

Caution therefore needs to be taken in terms of all industrial activity in the Arctic. This necessary caution is harder to secure in the Arctic because of the more extreme weather conditions: higher winds, more frost and drifting pack ice. Therefore it is of utmost importance that environmental protection procedures are put in place with every new project in the Arctic.

Many nations, like Norway and Iceland, are somewhat worried that the Russians, who own by far the most natural resources in the Arctic, will not set and adhere to stringent enough environmental standards. This holds especially true for standards regarding oil and gas drilling in Siberia and in the Barents Sea. The standards for the building of tankers will also have to be very high, not least because of high winds and the danger of sea ice.

In 1969 professor *Ian McHarg* published his book *Design with Nature*. This book laid the foundations for a design philosophy based on the approach that all design and planning needs to be based on knowledge about the natural environment and that this knowledge should be used to help adjust designs and plans to nature.

This methodology has slowly been gaining ground in the world but will in the future become even more important, since we cannot assume

Russia and Norway will increase their oil production in the Barents Sea, and export it by tankers to Europe and North America. About 500 tankers of 100,000 tons each are projected to pass Iceland yearly by 2015, causing great environmental worries.

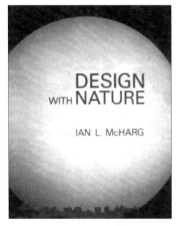

This book promotes the idea that every design should be based on knowledge about nature.

An overview of changes that come with global warming.

The matrix shows – in the first column – the main changes in weather that can take place with global warming.

Features coloured red are negative impacts of global warming, but features coloured green indicate positive impacts of global warming.

Some of the negative impacts will disappear as vegetation and habitation has adjusted to the new climatic conditions.

The half of the matrix that is on this page, shows the impact in the warm areas of the globe.

The opposite page shows the impact in the cold areas of the globe – where, indeed, many of the climate changes have a positive impact.

		In warm areas				
	Impact on →	Vegetation	Organisms	Land	Air	Water
Temperature	Warmer	÷ Bad if not more rain	Can seek colder areas	Warmer/dryer topsoil	Warmer winds Wildfires	Warmer water
	Heat waves	÷÷ Bad if not more rain	÷ Northern species suffer	— II —	÷ Turbulence Wildfires	— II —
	Colder	OK for northern plants	Can seek sunny areas	Colder/ wetter topsoil	Colder winds	Colder water
	Cold spells	Kills sourthern vegetation	Southern species suffer	— II —	— II —	— II —
Precipitation	More precipitation	More growth	Can seek shelter	÷ More water erosion	More moisture Cloudier	÷ More floods
	More extreme precipitation	÷ Erosion Plant suffocation	Seek shelter Avoid floods	÷÷ Far more water erosion	— II —	÷÷ More extreme floods
	Less precipitation	Less growth	Can seek wetter areas	Dryer topsoil	Dryer winds wild fires	Lower water level
	More extreme droughts	Dryes up – dies	Seek secure waterholes	— II —	— II —	÷÷ Disappears in some places
Airflow	Windier	Less growth Dry up	Can seek shelter	÷ More wind erosion	More turbulence	Bigger waves
	Stronger winds	÷ Braek plants	Can seek shelter	÷÷ More wind erosion	— II —	÷ More extreme waves
	Calmer	More growth	More comfortable	Less wind erosion	Less turbulences	Smaller waves
	More calm spells	— II —	Micro-climate too hot	— II —	— II —	— II —

In cold areas

Vegetation	Organisms	Land	Air	Water	Snow/ice
++ Good if not less rain	+ Good for most	+ Warmer topsoil	+ Warmer winds	+ Warmer water	+ Less snow/ice glaciers recede
÷ Bad for northern species	÷ Northern species suffer	+ Warmer topsoil	+ Warmer winds	+ Warmer water	÷ Sudden runoffs Floods
Bad for southern species	Bad for southern species	÷ Colder/wetter topsoil	Colder winds	÷ Water freezes Frost damage	÷ More snow/ice Glaciers expand
Kills southern vegetation	÷ Southern species suffer	÷ Colder/wetter topsoil	— II —	÷ Water freezes Lack of water	Less runoff Fewer floods
+ More growth	Can seek shelter	More erosion	More moisture Cloudier	More flooding	÷ More snow in cold periods
÷ Erosion Suffication	Seek shelter Avoid floods	÷ Far more erosion	— II —	÷ More extreme floods	÷ More extreme snowloads
Less growth	Can seek wetter areas	Dryer topsoil	Dryer winds	Lower water level	Less snow Glaciers recede
÷ Dries up Dies	Seek secure waterholes	— II —	— II —	÷ Disappears in some places	— II —
÷ Less growth Dry up	÷ Bad for southern species	÷ More wind erosion	÷ Higher windchill	More waves	÷ Snow less evenly distributed
÷ Braek plants	÷ Some die	÷ More wind erosion	÷÷ Higher wind-chill factor	÷ Bigger waves	÷ Snow less evenly distributed
+ More growth	+ More comfortable	Less wind erosion	+ More comfortable	+ Smaller waves	Snow more evenly distributed
+ More growth	+ More comfortable	— II —	+ More comfortable	+ Smaller waves	— II —

This right page: Impact of warming in cold areas.

The + and ÷ indicate where the most positive and most negative impacts of the climate changes are already being felt, and will be felt still more.

These are generalizations, and the impacts vary, e.g. because of the influence of other factors.

There are books that deal with the subject of each of the 132 boxes in this matrix. What we primarily realize, looking at the matrix, is that the impact of global warming is very multi-faceted.

We also see – to our amazement – from this matrix, is that in cold areas the impacts of global warming are within some of the most important areas positive.

To Fight Change

Man thinks:
I can control Nature!
And man wishes:
May nothing ever change!

On a small scale
Man is in control.
But in mega-events
He needs to reconsider.

Nations build levees
And gain land.
The fallacy becomes clear
With rising sea levels.

Not to fight the oceans
Is the lesson here.
The oceans will rise –
It is ours to yield.

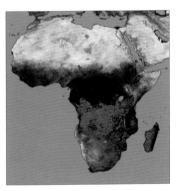

The red on the map shows areas of most increased wildfires in Africa.

that the standards and conditions that we are used to will remain adequate or optimal for future conditions. We have the advantage of being able to study and plan for large areas. With the help of the "design-with-nature" planning process we are able to spell out in what way facilities can most beneficially be placed in an environment to assure little or no environmental damage. What will make this methodology especially important is that we can rather easily introduce into the process any expected or measured changed environmental features that come with global warming.

On the previous spread there is a matrix that provides an overview of how changes in some basic environmental features, like temperature, precipitation and airflow, have an impact on vegetation, organisms, land, air and water. The left hand page of this matrix shows the impact of these projected weather changes in the warm areas of the globe. The right hand page shows the impact in the now cold areas of the globe.

A general characteristic is that most of the climatic changes that will come with global warming will have a negative impact in warm areas but a positive impact in cold areas. As the matrix shows, these generalizations do not always hold true and they, of course, differ from one region to another. The temperature changes are shown in the matrix, both as a change towards a warmer climate and towards a colder climate. This is shown because even though, in general, most of the world is going to get warmer there are areas that will get somewhat colder.

In some cases, extremes in weather can mean colder cold spells than earlier. The same holds true for precipitation because although, in general, where there will be more precipitation, there may also be some regions, or periods of time, that will experience less precipitation.

The main finding of the matrix, as concerns the warm areas of the globe, is that the increasing heat waves will be very bad for vegetation, especially in areas that will not get more rain. Heat waves and air turbulence will also increase the number of wildfires, which is something that has already started to happen in the world. In the USA, for instance, the average area that has burnt

in one year has more than doubled in the time span from 1970 to 2000. This increased burning coincides with the climatic warming in the regions in question.

A very special negative feature that is not mentioned in the matrix is more insect outbreaks. In recent decades several such instances have been registered. Birch forests in Scandinavia, for instance, have of late been destroyed by moths and forests in Yukon in Canada by spruce beetles. The influx of new pests that come with global warming will also, in some areas, affect the health of humans in a negative way.

Residents of poorer countries will be more at risk in these outbreaks because of fewer resources and less knowledge about how to deal with them. Increased environmental stress on existing animal populations will also create conditions for diseases that can be transmitted to humans, such as the West Nile virus. The West Nile virus has, for instance, spread enormously in Canada in the last decade.

All these threats and dangers are a cause for alarm and nations have to prepare themselves for meeting these challenges. The problems will be larger in areas where the climatic changes will be most extensive, primarily in the Arctic, and also in areas that are already environmentally vulnerable because of extreme heat, dryness and wind.

Scientists are able to tell us in which direction deserts will be spreading, they can tell us where there will be more floods and where there probably will be higher winds and hurricanes. These extremes in weather have been hitting the world at increasing rates in the last decade; major 100 year floods coming every two or three years in some areas in Europe, recently. Because of the increased warmth of the seas – for instance in the Caribbean and the Gulf of Mexico and the China Sea – the number of hurricanes and their force have been increasing, causing extensive damage in the south central USA and elsewhere.

Best known are the terrible impacts of floods as the levees constraining the Mississippi River gave way in 2005, under the impact of Hurricane Katrina, and large areas of New Orleans were flooded.

3 Changes in Human Habitats

It is quite logical that the study of the impact of climatic changes started with a study of climate itself and then continued with the impact of the climatic changes on natural environments. The first two sections of this book dealt with these rather well studied subjects. The last section then added an illustration of how certain changes in natural environments lead to changes in human habitats because even in today's technological societies, human activity and habitats have a close connection to many natural features.

The main patterns of human habitats today still follow, in a fundamental manner, the patterns of agriculture, fisheries, industry, tourism, etc. *Future changes in these occupational patterns* will therefore eventually also *lead to changes in human settlement patterns.*

As earlier described, the world's climate will, generally speaking, become warmer, wetter and windier with global warming. The impact of these climate changes on the various forms of habitation on Earth will be quite varied and differ according to climate zones. The matrix on the next spread is therefore divided into three: *Tropics, Temperate areas* and *Polar areas* in the columns of the matrix, with the main changes in terms of Agriculture, Fisheries, Tourism etc. explained for each of these climate zones. In the tropics and in Earth's dry, central areas, agriculture will suffer, leading to problems in adjustment of human settlement patterns and type of abodes. In the polar areas, on the other hand, the fine vegetation that grows now in mid-Scandinavia will start to develop further north. This has already started to happen and eventually substantially more human habitation will develop in some of the sub-polar areas.

The increased power of such climatic factors as rain and wind will make natural processes more dominant in the future. When natural processes reach such strength that man has a hard time coping with them, we talk about them as *hazards.* In cases where human habitats are not threatened, we are willing to understand that it is a basic characteristic of nature to be dynamic and ever-changing. Therefore, for instance, we think of coastal erosion as a natural process but not a hazard, if it is in a totally unsettled area, whereas if a coastal settlement is threatened by erosion, we start to call it a hazard.

Earlier cultures were more cautious, and thus more sensitive to what types of areas might be in danger because of natural processes. Modern man's *haughty confidence in his own ability,* on the other hand, has in many cases reached a level of arrogance and overbearing towards dangers in nature with the result that we today often disregard the potential power of natural processes. Many houses in southwestern California, for example, have been built in areas prone to earthquakes and landslides. In general, very many settlements and constructions of today are not designed to meet extremes of natural processes. China is an ancient civilization that has, for thousands of years, been co-habiting with its large rivers and their floods. The builders of the dense settlements along those rivers had no other choice than to design the settlements in harmony with the fluctuations in the rivers. Only in modern times has China, like other modern societies, interfered with the natural flood plains and built in a way that cannot stand the test of time.

In spite of the highly advanced ancient knowledge of the Chinese about *environmental management* – which modern man thinks is a new invention – occasionally in the history of China the most extreme river floods have surpassed the limits of the management systems, causing enormous casualties. Modern technology, however, provides us with techniques that can obliterate most of these problems of flooding, mostly by building dams that are capable of holding back floods of far larger dimensions than earlier.

Many people think that the huge dams in the *Three Gorges* area are being built solely for the generation of electricity. The expected flood control, however, is no less important and now, as the Chinese are increasingly modernizing, more sud-

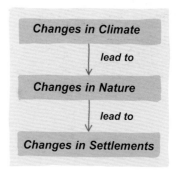

Logical sequence in the changes that come with global warming.

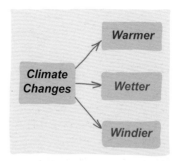

Changes in weather that come with global warming in the North.

Impact of global warming on various human activity.

The main changes that come with global warming are that the climate becomes warmer, wetter and windier – mostly in the North.

Even if warmer, wetter and windier holds true, in general, certain areas can become colder, dryer and less windy. Also, these weather patterns can occur in different combinations.

Increased energy in weather patterns will lead to an increase in various natural hazards, as the lower part of the matrix shows.

Increased hazards along coasts, will mostly be caused by the rise of sea level. Many predict a rise of 50–90 cm in this century – and many additional metres in the centuries to come, if the warming continues.

		Basic industries				
	Impact on →	Agriculture	Fisheries	Tourism	Energy	Water
Warmer – Wetter – Windier	Tropics	++ Less water Spreading of deserts	+ Increased production of the sea	÷ Too hot in some areas	From the sun	÷÷ Problems in dry areas Evaporation
	Temperate areas	+ More growth if enough water	+ Increased production of the sea	More comfortable	Sources as today	÷ Lack of clean water
	Polar areas	++ Same biozone as now in central Scandinavia	++ Increased production of the sea	+ New tourist areas	+ Hydropower in ice free rivers	++ Abundance of clean water
Increased natural hazards	Coastal floods	Aricultural land on coasts reduced	—	÷ Coastal facilities threatened	—	÷ Salt pollution reaches into aquifers
	River floods	÷÷ River floods hard to control	—	÷ Facilities by rivers threatened	—	Pollution of wetlands and aquifers
	Landslides and avalanches	÷ Threat in steep terrain	—	Bad for mountain tourism	—	—
	Deserti-fication	÷÷ Agricultural land reduced	—	Less tourism in deserts	—	÷ Less water
	Wild-fires	÷÷ Increase in dry and warm areas	—	Smoke means inconvenience	Can threaten facilities	Less water more fires

Transportation				Vulnerable areas	
Land	Sea	Rivers	Air	Low coasts	Permafrost
—	—	—	—	÷ Coastal erosion Areas flood with rise in sea level	—
—	—	—	—	÷ More coastal erosion	—
÷ Less frost: less suitable transport	++ Less ice: easier shipping (long-term)	÷ Less ice: easier shipping (long-term)	÷ More wind: bad for airplanes	÷÷ Less ice, more wind: coastal erosion	÷ Thawing of soil: foundations weakened
Coastal roads may disappear	Harbours damaged	—	÷ Bad for airports on low-lying coasts	÷ Coastal settlements suffer	÷ Thawing of soil: more coastal erosion
Temporary disturbances	—	Temporary disturbances	Bad for airports on flood plains	River floods: erosion	÷ Thawing of soil: more river erosion
Can block roads	—	Can dam up rivers: flooding	Bad for airports in landslide areas	Landslides by steep coasts	Thawing of soil: more landslides
Driving through desert areas difficult	—	—	Some airports threatened	Less desertification at coasts	Erosion: deserts in cold/dry areas
Fires can block traffic	—	Rivers: lines of defence	Temporary disturbances	Fewer wildfires along coasts	Forests topple: dead wood burns easily

Of most impact by global warming is ...

... the spreading of deserts, reduction of sea ice, thawing of permafrost and higher sea level in the long run.

Less sea ice means that shipping lanes between the North Atlantic and North Pacific will open.

Poor countries in low-lying areas, and countries with short-sighted environmental planning will suffer most from the rise in sea level.

Countries with much land, however, can retreat from coasts and rich countries can afford to build coastal defences.

*The permafrost areas of the polar regions will suffer from the higher temperatures.
In the permafrost areas settlements, roads, airports, etc. are constructed on frozen soils, but not on solid foundations.*

With less frost foundations are weakened and the constructions tend to settle unevenly and to crack. As the permafrost disappears the surface soil becomes mud at first.

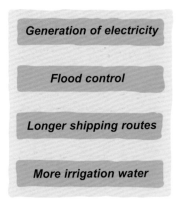

Generation of electricity

Flood control

Longer shipping routes

More irrigation water

Gains from damming rivers. The generation of electricity is not the only gain!

... Of course there are also many negative impacts of dams.

With the rise in sea–level coastal areas need to redesigned and some of them abandoned.

den floods to be expected there, dams for controling the floods become more important.

The third gain acquired by these dams will be that the higher water level behind the dams will allow larger ships to go further inland. The reach of 10,000 ton ships inland will, for instance, increase by some 600 kilometres in the Yangtze River. As shipping is very environmentally friendly compared to road traffic, the regions in question will experience huge environmental gains from this. The dam also means that water in floods is not lost out to sea, so water can be channeled north to the dry Beijing region.

The lower part of the *matrix* on the previous spread describes the impact of, for instance, *coastal and ocean flooding hazards* on various basic industries. Other hazards dealt with in the matrix include: an increase in *landslides and avalanches, desertification and wildfires.* The right hand side of the matrix deals with the impact of these hazards on transportation on *land, sea, rivers* and *in the air.* Finally, there are two columns that describe the impact of these hazards on two of the most vulnerable areas on Earth: *low coastal areas* and the *permafrost areas* in the polar regions.

Some of the natural processes described here will mean that some urban areas of the globe will become much less habitable. On the other hand, we also need to keep in mind that – especially in countries that have sufficient space for development for human activity and human habitats – *there are areas that will almost be totally free from the impacts of natural hazards!*

In general, it can be said that with proper planning there need not be, or almost need not be, any natural processes that need to be regarded as *natural hazards in sparsely populated* countries. Today there are, of course, settlements that are placed in areas that are in considerable danger because of the increased occurrence of storms and strong winds and other natural processes. Governments therefore need to form policies and define on maps the areas that are in most danger of present and future hazards.

The more extreme forces of nature that are to be expect with global warming, in many cases, will mean a vast extension of earlier *hazard areas.* As an example of a necessary precaution, sensitive buildings should not be allowed to be built in areas that are now defined as "100 year river flood plains", as 100 year floods are now, in some places, occurring every five years.

Rising *sea level* and *increased coastal* erosion mean that flood prone areas along coasts also have to be redesigned and re-determined in view of the new realities. In many areas the natural coastal defences will not be able to withstand ocean forces with higher winds and higher sea levels, except at high and rocky costs. Therefore, a *coastal policy* also needs to be created that determines where more funds need to be allocated. Such a coastal policy should also determine in what instances it is advisable not to invest in any defences because the situation, practically speaking, is a lost battle. In such cases the rising sea level needs to be accorded space in the areas that the water will naturally claim.

In general, it is not a wise policy to try to fight mega-scale changes in nature along coasts; it is instead a much wiser policy *to retreat to higher ground* in order to insure a greater degree of safety for buildings, objects and humans. In studying the expected effects of global warming, we can slowly start to realise that our habitual ways of trying to control nature and combat it, have to be changed. Expected global changes in the natural environment *call for changed thinking and changed responses.* Today's haughtiness and urge to test the limits of how far people can go in terms of challenging the forces of nature will have to give way to more realistic approaches. Humankind – especially faced with the expected increased ferocity of weather patterns – needs to learn that the concept of nature includes a panoply of enormous forces that we will do well to respect if we are not to be too severely devastated by catastrophes.

This means, in terms of design and planning, that humankind has to adapt to *a different planning philosophy.* The study of the natural climatic patterns of the past as a means to create reliable standards on which to base future construction and planning is no longer enough. It is not only that the power and intensity of patterns like increased wind and rain that will change but also that these patterns will – in terms of seasons and regions – be more unpredictable.

The *design philosophy of the future* therefore needs to be characterised by a policy that seeks to make all plans and all designs able to cope with a very wide variety of occasions. In terms of architecture and construction this means, for instance, that buildings and other construction should be *simple, robust* and *resilient*. This is quite a change from the design philosophy of today, where many architects and engineers go out on a limb in terms of challenging the forces of nature.

Again the main characteristic of the needed philosophy is *humbleness* in terms of our ability to build something that lasts, humbleness in terms of respecting the forces of nature, and humbleness in terms of admitting that we cannot use our empirical knowledge of nature on which to base construction and planning standards.

The new philosophy must proclaim that we should always try *to seek the most secure areas* and always try to make all construction as robust and resilient as possible, because all building and infrastructure will have to be able to withstand forces far greater than today's accepted standards require.

The world has already started to experience extensive alternations in weather patterns. For instance, in the winter of 2005 to 2006 some areas in Europe experienced much more precipitation, in this case snow, in a shorter period than all standards assume. This meant that the extreme snow loads broke roofs all over central Europe, for instance in Austria, Poland and Moscow, killing several hundred people in the process.

In Scandinavia more than usually extreme winds tore off roofs, downed trees and closed bridges for a much longer time than has been habitual in recent decades. Increased sudden rainfall in southern England, France, Germany and Austria and other countries has, in the last few years, also meant many more floods than previous statistics have led us to expect.

In the spring of 2004 central Europe, for instance, experienced a so-called 100 year flood – supposedly only occurring once a century, but only two years later there was another 100 year flood. To make matters worse, the river channels have become narrower. This means that the flat areas; the flood plains, that were earlier available for the spread of the flood waters, have been closed off and filled with construction so that the only way the increased water can go is upwards, resulting in a rise of some 7 to 9 metres in some areas.

The damage caused by floods is more than people realize. The floods in 2005 in the Czech Republic, for instance, caused more damage than all the damage caused in the Second World War.

In the case of river flooding we not only have to deal with problems of more extreme rainfall in short periods of time and the narrower river channels, but today's deforestation and expansion of hard surfaces that are created by streets, roofs and parking lots, also contribute greatly to the problem. The hard surfaces mean that the time it takes rainwater to reach rivers has become much shorter, largely because earlier soil and vegetation used to slow the rush of the water, both through absorption and the delaying effect of the varied surfaces.

1) Construction doesn't last

2) Need to be humble in respect of nature

3) We can't rely on past experiences alone

These three realizations form a basis for the needed, new design philosophy.

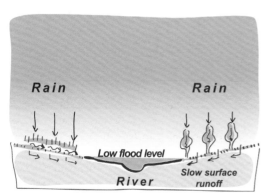

Rivers in Nature: Vegetation, absorbing surfaces and few obstacles mean lower flood levels.

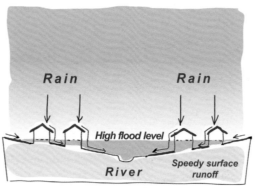

Rivers in Cities: Obstacles and speedy runoff over closed and hard surfaces mean high floods.

Floods are often, falsely, blamed on global warming alone.

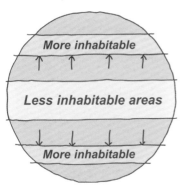

Global warming means that the warm central regions of the globe will become less inhabitable ...

... but instead, the sub-polar areas will become more inhabitable.

We have been considering the impact of more precipitation and more sudden bursts of precipitation. There are in addition hazards linked to less precipitation that will also occur in some regions of Earth, even though in general total precipitation is expected to increase. The problems associated with less precipitation will mean increased heat and evaporation.

The problem of less precipitation will be most serious close to the deserts, which will expand, for instance in the sub-Saharan region. This is a region where desertification is already on the increase, not only because the climate is getting drier but also because of erosion caused by over-grazing and too much harvesting of vegetation for fuel and other uses.

Obviously, water is a prime requirement in agriculture, as for all types of vegetation, so some areas of the world that will experience less rain and more droughts may become dust bowls. In such areas the area of land suitable for agriculture will be reduced, along with agricultural yields.

In the areas of considerable vegetation, a reduction in precipitation and increased heat also means a higher risk of wildfires. In the summer of 2005 the wild fires in Portugal got so much out of hand that the government had to seek assistance from the European Union to cope with the problem. Water, at the same time, became so scarce that hotel owners were not allowed to fill their swimming pools. Less precipitation in Southern France of late has meant that water-demanding crops like maize can hardly be raised there any more.

Again, it is *environmental management* and *environmental planning* that are the main tools that can help us to cope with these problems. To help control wildfires, for instance, forests need to be cut by fire breaks. Forests or wooded areas close to settled areas have to be cleared and lakes need to be created to help create buffer zones between settlements and forests.

In terms of possible meassures we again, need to keep our *global overview* in balance. Let's for instance keep in mind that there are areas that today have more than enough water; areas that people in drought-stricken areas can move to.

A balance score card will tell us that areas that will be lost – for instance to desertification and to too much warmth and dryness – *will be substituted and "paid for"* by areas that are now uninhabitable in the sub-Polar regions, areas that will become fruitful and inhabitable. The pattern of climate change, in general, is that the central areas of Earth will get most of the negative impacts, whereas the positive will be felt more towards the poles.

Again we need to remind ourselves that we need to adopt a *philosophy of humbleness*. This, practically speaking, means that we need to realize, and accept, that we will not be able to withstand or adapt to the changes as well as we have done in the past – and the changes, in certain areas, will be so great that there is no other way than to yield; to move out of the areas.

The huge changes in the natural patterns are making it clearer to us than ever before that it is the basic nature of Earth to be in *constant flux;* to be *dynamic.* If we try to maintain static political borders, static lifestyles, fixed areas of habitation, we are bound to be overrun by these changes. We need to prepare ourselves for accepting a new *paradigm of dynamism* where all structures and all organizations, as well as political and social systems, need to allow for a more flexible and dynamic way of living and choosing an area for habitation.

This new paradigm for instance means, that political borders that limit migration need to be abolished, and that laws on ownership of land have to be changed.

Political borders

Static lifestyles

Fixed habitations

Cultural barriers

Language barriers

Static features that prevent people moving from less habitable areas to more promising areas.

Deserts have spread earlier in periods of drought and warming. The picture shows a sand storm cloude approching a farm.

4 Three Strategies to Global Warming

Before we start in this chapter to study three strategies that are possible as a means to respond to the problem of a warming world climate, it is necessary to place this particular environmental problem of global warming into the wider context of all the main global problems that exist in the world today.

In 2004 the *EIA Institute* in Denmark called for a conference in Copenhagen where the then most threatening global problems – as well as the most sensible solutions to these problems – were discussed. Pro and contra speakers introduced the ten main global problems or challenges.

One of the challenges discussed in the conference was global warming and the opportunities that are at hand to combat it. The three opportunities that were discussed were: the measures of the Kyoto Protocol, optimal carbon tax, and finally, value at risk carbon tax.

To propose priorities among the proposals for confronting the global challenges, a panel of economic experts comprising eight of the world's most distinguished economists were invited to consider the issues. The panel was asked to answer the question "What would be the best way of advancing global welfare, and particularly the welfare of developing countries, supposing that an additional $ 50 bn of resources were at government disposal?"

The ranking of the opportunities to combat the global challenges – as judged by the panel – is shown in the table below. First in rank is the challenge of contagious diseases and the highly effective opportunity of controlling HIV/AIDS. The second aspect in the ranking also offers a rather easy solution, that is the opportunities provided

		Challenge	Opportunity
Very Good	1	Communicable diseases	Control of HIV/AIDS
	2	Malnutrition and hunger	Providing micronutrients
	3	Subsidies and trade	Trade liberalisation
	4	Communicable diseases	Control of malaria
Good	5	Malnutrition and hunger	Development of new agricultural technologies
	6	Sanitation and water	Community-managed water supply and sanitation
	7	Sanitation and water	Small-scale water technology for livelihoods
	8	Sanitation and water	Research on water productivity in food production
	9	Governance and corruption	Lowering the cost of starting a new business
Fair	10	Migration	Lowering barriers to migration for skilled workers
	11	Malnutrition and hunger	Improving infant and child nutrition
	12	Communicable diseases	Scaled-up basic health services
	13	Malnutrition and hunger	Reducing the prevalence of LBW
Bad	14	Migration	Guest worker programmes for the unskilled
	15	Climate change	Optimal carbon tax
	16	Climate change	The Kyoto Protocol
	17	Climate change	Value-at-risk carbon tax

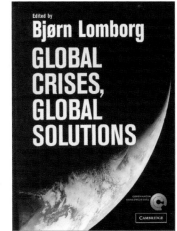

The table showes the final ranking of seventeen opportunities to combat ten major global problems. Global warming is one of these ten global problems. What comes as the biggest surprise is that the panel of experts ranked opportunities to combat global warming lowest in this table.

Global warming was one of ten global crises discussed and ranked at the Copenhagen conference.

Emissions:

USA	30.3%
Europe	27.7%
Russia	13.7%
Southeast Asia incl. China, India	12.2%
Central and South America	3.8%
Japan	3.7%
Middle East	2.6%
Africa	2.5%
Canada	2.3%
Australia	1.1%

The USA and Europe contribute, by far, most to global warming, but the development in many developing countries is very fast.

by trade liberalization to combat the problems of hunger and today's trade-barriers.

Surprisingly the three policy options discussed at the conference *to combat climate change* got *the lowest rating*. A summary text by V. Smith offers an evaluation of the expert panel view: "It is clear from both the science and economics of intervention that those on earth who care about environment are not well advised to favour initiating a costly attempt to reduce greenhouse gasses (ghgs) build up in the atmosphere in the near future, based on the available information." In spite of this blunt statement it is, however, stated "...that the science has produced a consensus: global warming and global climate over the last century or so is attributable to anthropogenic releases of ghgs..."

The main economic argument supporting the opinion is that investments in curbing global warming will be very costly, is that it will probably not be felt until after many decades, and possibly also, that the warming be get so great anyway that

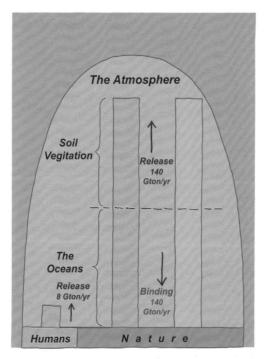

There is a balance of release and binding of CO$_2$ in nature, but nature has a hard time absorbing all the emissions created by humans.

human attempts to curb the emissions would do little to reduce the warming rate.

Let us now look at the picture below that explains the *global balance of release and binding of carbon dioxide (CO$_2$)*. For most part of Earth's history there was a balance between release and binding: The release of CO$_2$ by decay, burning or oxygenation, was absorbed by living organisms that produced oxygen.

The two main actors in this equation were the life-systems of the oceans and of living nature on land. Within both of these actors there is almost a balance; soil and vegetation on land releases, and receives, about 75 gigatons of CO$_2$, the oceans somewhat less, or about 65 gigatons in a year, together making 140 gigatons per year.

The CO$_2$ balance started to change with the advancement of industrialization around 1800, mostly because the power plants were mainly powered and the cities heated by dirty coal. The problem with this is the huge release of CO$_2$ into the air, that humans have no mechanisms to absorb to the emissions to maintain the balance.

At the beginning of industrialization, nature provided some possibilities for absorbing more CO$_2$ but on that absorbing side of the equation humans unfortunately increased deforestation, which hugely reduced the global capacity of land-based vegetation to absorb the extra exhaust. Today the anthropogenic release of CO$_2$ comes to about 7 to 8 gigatons per year. The main means people have at hand to help absorb these carbons is to reduce the cutting of forests and increase reforestation.

The most discussed strategy to curb greenhouse gas emissions is *the qoutas mechanisms of the Kyoto Protocol*. The Kyoto Protocol tryes to organize this qoutas mechanism with an international treaty. Many nations have already subscribed to certain percentages in reductions – but most developing nations are allowed an increase in emissions.

The problem with the Kyoto agreement is that even though 191 nations have signed it, the total emissions of these nations only constitute 1/4 of the human emissions in the world. It is, however, a conscious decision of the leading nations not to put pressure on the now industrializing nations to enter the agreement until after 2012.

The fairness argument behind this is that the old industrial nations have, for a long time, had a free hand in building up their industries without being limited by quotas, so the now industrializing nations should be allowed to be free of such restrictions in the first periods of the protocol. The main problem, in terms of getting this international agreement to be effective, is that some of the largest industrial nations, like the *USA, China* and *Japan*, have not signed the agreement.

Most pressure has been put on the USA to enter the agreement since the USA produces about 30% of world emissions, even if its population is only about 4.6% of the world population. In the latest Kyoto Conference in Montreal in November 2005, the USA together with a group of other Asian-Pacific nations, declared that they had a different program and a different approach to solving the problem.

The nations of the new partnership, called *A6*, are the *USA, China, Japan, India, South Korea* and *Australia*. In January 2006 they had a conference in Sydney, Australia, where they outlined their program for reduction of greenhouse gasses.

They divided their program into *eight tasks:* 1. cleaner fossil energy; 2. renewable energy; 3. power generation and transmission; 4. steel; 5. aluminium; 6. cement; 7. coal mining; and 8. buildings and appliances task force.

This new *Asia-Pacific Partnership on Clean Development and Climate* summarized its action-oriented world plan *in six points:* 1. coal gasification and other clean coal technologies; 2. extending the use of renewable energy; 3. improve the efficiency and reliability of electric power-systems; 4. deploy advanced manufacturing processes for cleaner aluminium and cement production; 5. adoption of building- and appliance efficiency standards, and; 6. capturing and using coal methane as a clean energy resource.

The main characteristic of this *A6 strategy* is not to impose any quotas on any of these nations but rather to help, in a co-ordinated manner, to develop new technologies for improving environmental and efficiency standards.

Many of the critics declared that the new partnership was merely a fig leave to cover the embarrassment of George Bush and John Howard, the only Western leaders who rejected the commitments of the Kyoto Conference in 1997. The Austrian Minister of Industry Ian Macfarelane replied to this criticism "The reality is that new technologies will deliver three times the savings in greenhouse gasses as the Kyoto Protocol will."

It is very hard, at this point in time, to make a judgement about how efficient the programs of the A6 will be, since there are no obligatory binding targets that these nations have placed on themselves. People who are sceptical of Kyoto counteract the critiques by pointing out that even though there are standards that have been agreed upon within Kyoto, only very few of the signatory nations have met the standards.

As it comes to the assessment of what will happen with the Kyoto and A6 strategies to combat global warming, it is almost unthinkable that the A6 nations, which constitute about half of the world's emissions, will join the Kyoto Protocol for the next period, starting in 2012. That might mean a break up of the protocol strategy altogether.

Viewing these matters in a very down to earth and practical way, it seems obvious that as long as the fossil fuels are inexpensive, the incentive to reduce the use of them will not be great. High fossil fuels are therefore needed to make alternative energy economical, which is essential to acchiese a reduction in emissions.

The big problem with the so-called "clean" energy alternatives is that they are not only costly but are also harmful to the environment. The large spaces needed for the solar panels and the visual and noise pollution produced by wind-power generation is, for example, very negative. Together, these renewables constitute only about 1.5% of the world's energy today and to increase this percentage substantially will be very hard.

Hydropower has some potential, but today it only accounts for about 7% of the energy share, and environmentalists in, for instance, south America and Africa, where a large potential for hydro power is located – are working very much against this type of power generation.

Nuclear power accounts today for about 6% of world energy production. Earlier the nuclear

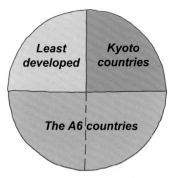

Devison of the world emissions: The Kyoto countries account for ¼. The least developed countries for ¼, and the A6 nations for ½.

Eight task forces
1. *Cleaner fossil energy*
2. *Renewable energy*
3. *Power generation and transmission*
4. *Steel*
5. *Aluminium*
6. *Cement*
7. *Coal mining*
8. *Buildings and appliances*

Action-oriented plan
1. *Coal gasification and other clean coal technologies*
2. *Extending the use of renewables*
3. *Improve efficiency and reliability of electric power-systems*
4. *Deploy advanced manufacturing processes for cleaner aluminium and cement production*
5. *Adoption of building and appliance efficiency standards*
6. *Capturing and using coal methane as a clean energy resource*

The boxes show the main policies of the A6 nations. They don't use quotas as the Kyoto nations do.

Most new nuclear plants have become safer, but if accidents occur they can be catastrophic. Little CO_2 emissions make them fesible to many countries. The picture shows vapor coming out of cooling towers.

Environmentally friendly machinery is luxury that poor countries often can't afford.

power plants were unsafe, as the Chernobyl disaster of 1986 showed us. Today most nuclear plants are considered very safe, but if an accident happens, the consequences will be huge.

In spite of this, objections to nuclear energy are on the decline. One of the main reasons why the developing nations do not want to see this sector expanded, is that new, unstable nations that are working in the field, may eventually be able to build nuclear weapons.

The amount of gas in the energy share is today about 24%, and is increasing. The problem with gas is that today it can primarily be economically transported through pipes, whereas oil can be transported via ships to almost every desirable location. But the liquefaction of gas is, however, picking up.

Petroleum account today for about 38% of the energy share but is predicted to reduce. *Coal* comprises today about 24% of the world energy share and is projected to rise, mostly because there is an abundance of coal available in developing countries, for example in China and India. Emissions from coal power plants are very high, but methods are being sought to reduce the emissions, for example by gasification.

It is a natural and instinctive response in the discussion of environmental problems facing the world today that it would be better to turn back the wheel of time and reduce the industrial production of countries, because it is primarily the increased production – that is in some relation with the increase in population – that is primarily leading to the increased environmental problems.

The problem with this thinking is that improved wealth of the developing nations is needed to combat environmental problems, because new technologies and measures to amend such environmental problems are costly.

Developing nations that would not be allowed to increase industrial production would probably permanently, be placed at a survival stage, so that they would not have the extra funds needed for financing the clean technology and other features it takes to reduce environmental problems.

Speeding up development and the increase of the wealth of nations is essential so they are able to deal with environmental problems. What fol-

lows naturally is that the main environmental problem in the world *is the poverty of the developing nations* because most of these nations don't have the funds to buy environmental friendly technology and install systems to operate the societies in an environmentally friendly way.

If the Kyoto nations succeed in imposing quotas and restrictions on the development of the poor countries, these countries will probably never get out of the poverty trap, and never reach the stage of a sustainable, environmental friendly society that we all want to see.

There seems to be a simple, common sense conclusion that can be reached from what has described here: *The Kyoto idea* of putting limits to emissions will probably not work as a strategy, and secondly: *The Asia-Pacific partnership nations* are probably going to expand their industries enormously in the future – and probably also their emissions. And the third conclusion: The poorest counties of the rest of the world will – first of all – probably not be able to join the Kyoto agreement, and secondly, they will probably not have the means to develop their economies in the way the vigorous, six Asia-Pacific states are planning to do. The general conclusion of this is that it is likely *that the idea of reducing greenhouse gas emissions in the world will probably not work.*

Earlier it was described how there is a balance between land-based nature and the natural systems of the oceans in terms of the release and absorption of CO_2 ... and that androgenic activity has put a stress on this balance.

In addition we discover, by looking at the overall picture of the globe, that certain occurrences in the natural environment will influence the overall CO_2 balance in various ways in the future. First of all, with increased warmth and rain, land vegetation will grow more and thus also its absorption of CO_2. On the negative side, it is to be expected that desertification will increase, exposing soils to oxidation with an increased production of CO_2.

Perhaps of by far the greatest concern is *the thawing of the permafrost* in the polar areas. These "permanently" frozen soils include huge piles of organisms that lived a long time ago – but have been prevented from rotting by the frost.

As the permafrost becomes more and more ice free, these piles of organic soils will start to rott ever more. It is now being attempted to calculate how much the resulting emissions will be. It seems that it will be in such quantities that it, in due time, it may possibly exceed all human emissions of the Earth today.

If this is really what is going to happen, it becomes less sensible to put costly restrictions on human activity for instace by qoutas, in order to reduced human induced emissions. The conclusion of the author of this book is therefore that the idea of stopping global warming is, sadly, probably not going to work, so that the climate of Earth is, almost certainly, going to get much warmer than now predicted.

This will mean that today's prediction of, for instance, the 4 to 7°C increase in temperature over land areas in the Artic by 2100, will fall short of the actual, and secondly, that the temperatures will continue to rise further after that.

If this turns out to be the way things happen, *the third strategy* man has available as a response to a warming climate – *to retreat* from warm and dry areas into cooler and wetter areas, will be the only practical strategy available – and this is the strategy outlined in this book. It is important to keep in mind that this is the strategy that organisms and humans have always used in global warming periods in history. And in a broader sense, this is *the philosophy of going with changes,* and *not trying to combat them.*

This, quite possibly, will be the only realistic strategy left for mankind. This book will outline this strategy and describe how this will physically appear in terms of the future settlement patterns on Earth. *The three most basic patterns will be:* 1) Migration of people towards cool coasts, 2) To migrate to higher and cooler areas, and 3) – which is the main option if the Earth's climate is going to become extensively hot – *to move towards the polar regions.*

The relevant political and social problems are many in our static structures of today so a *new parameter of dynamism and flexibility* has to be developed by the community of humankind. The problems associated with these migrating processes will be enormous; however, this is an option that is realizable, especially in the northern hemisphere where there is abundant space – space that today, is too cool for habitation, but with extensive global warming, will be areas of benign climate. These areas are mostly in Russia (Siberia), Northern Canada, Alaska, Greenland and Northern Scandinavia.

A positive aspect of this migration will be that the old inhabited spaces on Earth already have so many environmental problems that habitation in them, in some cases, has become, and will probably become, almost impossible. *The primary problems are:* 1) The reduced amount of water in aquifers and rivers, 2) Pollution of the soil and 3) Desertification. We have, so to speak, fouled our nest. The Arctic areas, on the other hand, have been protected by frost against human activity, so in most places soil, water and even air are clean. In addition there is in the Arctic an abundance of all of the main resources needed including oil, gas, coal, timber, water and clean agricultural soil.

Described in this way, it does not seem to need to lead to unsolvable problems to follow the strategy of migration. Migration is the way humans have, in all of the world's history, responded to global warming, or cooling; spreading towards the poles with warming and migrating away from the extending polar ice in periods of cooling.

There are certainly many obstacles for migration today, like fixed political borders and the unwillingness of people to accept other nations into their lands. *The solutions* are therefore not least *in the political and social realms* – and probably, most of all, *it takes a change of mindset.* We need to *accept dynamism* as a basic assumption of all functions on Earth, which means that we will have to accept *that change is a basic feature of human existence.* Humans have an inbred reluctance to accept the notion that a change is a must, people have a tendency to adhere to the norm, and to fear the unknown.

The global population has, in contrast to this, t*o develop an attitude that welcomes changes;* a culture that is happy that everything in the world is always changing, that every day and every new period is new and different, and that people and politicians will say "let us not try to conserve our today's ways of living and our today's environment,

---

The third strategy
to a warming climate consists of three patterns

1) Migration of people towards cool coasts

2) To higher and cooler areas

3) To move towards the polar regions

The primary problems
of the old inhabited spaces on Earth

1) The reduced amount of water in aquifers and rivers

2) Pollution of the soil

3) Desertification

Change: A Disaster?

Change is trouble,
Change is hardship.
But we can rephrase:
Change is challenge!

We can gain from change,
Or, let it defeat us.
The same with global warming:
We can gain from it.

There is a simple solution:
Yield to global warming,
Seek cooler areas,
Move toward the poles.

We have fouled our nest,
But new frontiers await,
Filled with opportunity
For a new beginning.

Controle of their fishing grounds is important to N European countries. The EU needs to relize that.

because it is of no use trying to do that in a world that has become so dynamic". The main and central idea of the philosophy of the future, and the slogan to cry out should be: *"Let's embrace change!"*

Planning and preparing for the changes is a very important activity. First of all we have to be able to outline, to map and describe, which will be the problem areas in the decades to come, and also which will be the desirable and habitable areas with increased temperatures in the world.

Every nation needs to create plans taking into account which areas need to be abandoned, for instance because of the rise in sea level, or because of desertification, and at the same time, point out areas where people will be safe from floods and expanding deserts and where there will be enough water and resources.

In many cases – mostly in the central regions of the world – such secure, resourceful areas will not be found within the borders of the state in question, so international cooperation on the relocation plans is needed. One of the first necessary steps is to abolish all political and national borders in respect of certain activities, as for example has happened with the establishment of the European Union.

Some of the Nordic countries of Europe will become more habitable in the future, like Greenland, Iceland and Norway, but unfortunately they are not members of the European Union, mostly because the EU has not been willing to recognize that the economies of these countries are so much dependant on their on the resouces of their continental shelfs and their fishing grounds that they can not risk giving up their soverainity over them. Looking, as Europeans, over the whole field of future northern migration, it is, however, Russia that, by far, holds the greatest potential in terms of future living-space.

Because of these important spaces in the north, the European Union needs to create a policy for the needed extension of the European space to the north. Such a policy is quite different to today's policy of extending only to the east into the former east-bloc regions.

Some Eastern European areas hold some promise, but many of them will be rather problematic in the future. Therefore it would be a wise policy of Western Europe, to open up a *Northern frontier* instead of today's Eastern one.

In Eastern Asia a northern migration will probably mean a revival of the Sino-Russian conflict at the boarder of China and Russia. In the nothern part of the Americas, the three main regions; Central America, Canada and the USA will have to be united in the EU way. If that happens there will be a free flow of people over the whole of the North American continent.

In the southern hemisphere, nations in the southern part of Africa will probably be out of harms way, like also in Southern America so people in these continents can migrate to there. However, migration to the Antarctica is not in the cards as scientists predict that the icecap there might even expand because of increased precipitation.

In the political scene racial tensions, and the divide between the rich and the poor nations of the world, is a problem that has to be addressed in order to allow this development to happen in an ordered way. Most social systems, like health systems and housing systems, will have to be designed and reconstructed to cover large areas – like in the European Union today.

With such changes, persons who wants to move north, will be able to get housing assistance and social and health services not only within their own national boundaries but also within the wider, new regional unions. Such unions need to be created in the various regions of the globe, as has happened with the EU in an exemplary manner in Western Europe, in the last half century.

II How Global Patterns Change

1 World View of Time and Change

Most of us realize that the way we think and act is directed by certain fundamental ideas in our cultures. Today people are very familiar with many aspects and differences between Western and Eastern modes of thinking. The governing *world view of the West* is characterized by ground rules which were mostly formed as a part of scientific reasoning in the 17th century by scientists like Newton, Bacon and Descartes. This world view is frequently called the Newtonian world view, where the nature of things is dealt with as the relationship of mass, force and speed. What stands out as lacking in this world view is that *time is not included* as an active parameter.

In physics, the strange absence of time, recognized by Einstein, led him to formulate his *Theory of Relativity* in 1905. The 20th century experienced further discoveries as concerns the active nature of time, as for instance in the work of Ilya Prigogine, regarding the activity of the *Arrow of Time,* in certain natural processes. As an example, if two liquids such as water and alcohol are brought together, they mix in the forward direction of time.

A further advancement in incorporating the Arrow of Time was a more proper understanding of the *Second Law of Thermodynamics*. This law explicates the active nature of time in such a way that every instance of work being carried out means that the amount of available energy within our solar system decreases. In scientific terms this is expressed as a tendency towards *entropy* or towards a state of less energy.

People might be inclined to view entropy as a negative force in nature, but in his book Order Out of Chaos Prigogine explains that life and creation depend on the dynamic interaction of entropy and order; "order through fluctuation" as Prigogine puts it. In his view all creation is to be seen as a process of this kind. This also translates into the understanding that everything that happens on Earth is a process, happens mostly in a fluctuation between the polarities of destruction to creation.

As we review these new findings of science and this approach to the natural forces around us, we start to understand better how t*oday's mechanical* and *static world view* is flawed and that time and change need to be included to form a new world view.

One of the main steps towards embracing *the new world view of time and change* is to try to remove the mental blocks of static time from our mindsets. If we do that, we can start to deal with the phenomena that develop from active time, phenomena like *change, development, dynamism, growth, cycles* and *processes*.

In the computer a simple rule in a chaotic process can create organic forms known in nature.

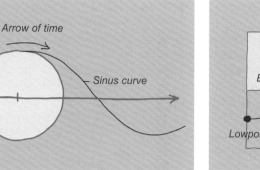

A cycle turning in the direction of time creates a sinus curve. Such cyclic curves are used to predict when occurrances hit highs and lows.

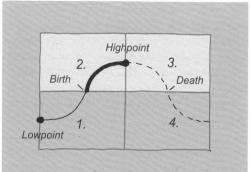

The cyclic process is divided into four sectors. Planning used to occure only in the 2nd sector but "life cycle design" embraces all four sectors.

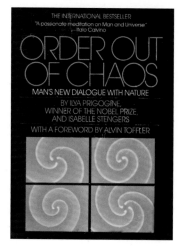

This book opens a new understanding of creativity where order emerges in a chaotic process.

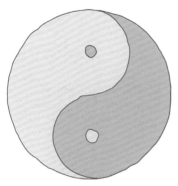

The curve symbolizes the dynamic interrelation of unlike features in Eastern thought.

One of the main authors of how new world views are created in science and in the world is Thomas Kuhn. In his book *The Structure of Scientific Revolutions,* Kuhn explains that in every given period there is a governing *normal science* characterized by certain ground rules, mostly stemming from physics. Occasionally a new world view starts to emerge through the works of scientists like the ones mentioned above. This Kuhn calls *revolutionary science.* Today the world is experiencing such a period of revolutionary science.

What emerges after a revolutionary period in science is a new era; a normal situation that is characterized by a new set of ground rules as, for instance, with today's addition of active time to physics. Kuhn calls this a *paradigm shift* and points outs, with many surprising examples, how difficult it is for the scientific community to accept that its basic assumption about the nature of the world needs to be changed. It is, not unsurprisingly, academia that fights the emerging revolutionary scientists most aggressively.

The new world view of time and chance will also, this time round, take a long time to accept. The dramatic situation of today as so many global patterns are changing, makes it an urgent necessarity to accept time and change as basic parameters of our existence. We can use writings such as Kuhn's book as a way to accustom ourselves to the new world view of time and change and its profound consequences.

One of the ways to acquaint ourselves with the new, emerging world view is to study Eastern philosophy and theories of revolutionary scientists, like Niels Bohr, as they explain how they have come to realize and develop their basic incentives for their new revolutionary scientific ideas. Bohr's main contribution was his description of the *complementary nature of the world* in a scientific way.

Eastern philosophy explains *complementarity* as being rooted in the dynamism involved in every subject in the world, characterized by the words *yin* and *yang.* Yang stands for the "manly" and aggressive features of the world and yin for the "feminine" and yielding qualities. In Eastern thought both features are considered equally important because neither could exist without the other. In the Western scheme of things, on the

other hand, such pairs have contrasting values; for instance the manly and aggressive values are commonly seen as admirable but the female and soft as insignificant. The *emancipation movement* of Western women has contributed greatly to the realization that not only masculine qualities and values are important but also the softer "feminine" values. Although, in the West, this debate has lasted more than a century, the softer feminine values in society, a value like humbleness and the value of being able to yield in appropriate situations, has a long way to go.

Because of what has now been described in the previous chapters, we understand that this book fundamentally deals with *man's relationship with Nature.* The book declares a "war" on the characteristic Western tendency not to respect the forces of Nature, but rather to try to control and manipulate them. This disrespectful view of Nature and the world is today being exposed more and more as a terrible and catastrophic way of understanding the world and working with Nature.

One of the best ways to start to get a more profound understanding of how wrong a premise today's Western arrogance is, is to study the ongoing clashes with Nature. These clashes will increase with the changing climatic patterns. It, however, can be said that the catastrophic occurrences linked to climate change bring one good thing with them, being a lesson that is teaching the whole world that our ways of *working against,* rather than *with Nature,* have been very wrong and that we, most definitely, need to change our ways.

The haughty Western attitude not only manifests itself in a very unwise attitude to Nature but also in an arrogant attitude towards many other things, such as towards the less powerful, both within societies and in international politics. We, for example, have a tendency to close our eyes to poverty and shut off concerns about what is happening in the third world. In today's world, where *globalization* has reached such a degree, environmentally, economically and politically, that no nation is an island anymore, these attitudes have to be changed. If, for instance, we do not help the developing countries to get out of today's poverty trap, they will not be able to deal with the envi-

The Structure of Scientific Revolutions
Second Edition, Enlarged

Thomas S. Kuhn

"A landmark in intellectual history."
—*Science*

The concept of active time helps us study social pheomena as a process, as in this book.

ronmental issues and will thus continue to pollute rivers, oceans and the air, resulting in an environmental depredation that will hit us all.

One of the basic features that result from the recognition of active time is *change*. It is one of the main tasks of this book to describe how everything on Earth and in our lives *is in a constant flux*. As we come to understand this, we will also start to understand that all changes follow patterns. It is therefore a main subject of the book to study *patterns;* to map them and create *pattern typologies,* to gain a more formalized understanding of what patterns there are and how they function.

Patterns of change have repeated themselves throughout history. Patterns in small spaces or time frames are not of interest in this book, but rather larger *megapattern trends,* which are spatial trends that extend over long spans of time. These time spans need to be long enough, and the one that is tracing or mapping the patterns must not to be influenced by occasional deviations caused for instance by short-term cyclical fluctuations.

The prognosis presented in this book is mostly based on a study of *megapatterns in Nature that come with climatic change*. Climatic change is a reality that has occurred many times before in the history of Earth, and has also happened before in historical times, so we can look at the history of the world and see megapatterns in how Nature and humankind have responded to such changes earlier.

Global warming, therefore, is not new and Nature and humans have found ways to adjust, even to enormous climatic changes, without suffering too much in the process. Our world today does not seem to include the right methods, social structures or frame of mind to take on global warming in a positive way. It is therefore an important task of this book to outline, through the study of earlier megapatterns, how we can adapt to such changes, largely by finding ways to get out of harm's way, mostry by retreating out of areas that become hazardous or in other ways unfit to human habitation. It is of high importance to learn to work with all these changes without viewing them as trauma and catastrophes, as the world of today does.

The most basic preamble for being able to put ourselves in the right frame of mind is not to try to combat the changes but rather to embrace them, since the climatic changes will be so enormous that there is no other way than to go with them in order to survive them – and surprisingly, to learn to benefit, and even enjoy them.

This way of talking about time and change will probably hit most readers of this book as being rather abstract. However, looking at the history of the world, we see that everything changes, even continents and organisms.

Today there is, of course, more acceleration of the warming of the world climate than we have been used to of late. However, looking at the history of the world, we see that such large climatic changes have often come before, and often very abruptly. Such catastrophic change has happened before, for instance because of a meteor impact that created a dust layer making the world has cooled very suddenly, so much so that terrestrial animals, dinosaurs and those small mammals which existed almost froze on their feet. Some early gigantic volcanic eruptions have also created a similar effect of cooling the climate. What is happening this time around is the reverse; the warming of the climate. The warming this time is caused by accumulation of greenhouse gasses in the atmosphere that trap the infrared warm radiation and prevent it from escaping into space.

It is a general characteristic of humans to live only in the moment, and not to have an overview of what is happening in the world around us. In this respect we are not very much different from the day-fly that sits in an ancient tree, chatting with a caterpillar. Says the caterpillar: "Do you know that this tree is a living organism?" "That can not be," says the day-fly, "I have been living here my whole life and I have not seen it grow a bit."

Another example of our poor understanding of time can be illustrated by the hands of the clock. If we do not think about the clock in an abstract way but rather in the way we are able to perceive it with our eyes, we would not believe that the hands of the clock are moving. This example helps us realize that we cannot rely on our perceptions for understanding the dynamism of the world; we have to turn to abstractions and scien-

Patterns of Change

From dust to dust;
The story of planets.
This same pattern
Applies to everything.

To a day-fly,
Not much changes.
The same holds true
For us humans.

Looking at history,
Continents change
Organisms change –
Nothing is static.

Periods of warming,
Periods of cooling –
They come and go:
Patterns of change.

The North American and Eurasian tectonic palates converge by 2 cm in a year. Iceland thus becomes 2 metres longer every 100 years.

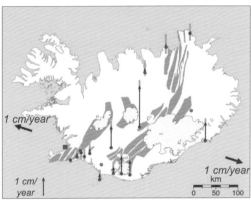

The vertical arrows show how much each part of Iceland is rising now, but the SE–corner for instance, will rise by about 2 metres in this century.

tific documentation of past developments, as well as instrumental measurements of the secular forces at work today, to understand the nature of the world.

In some areas of the globe it is easier to understand the dynamism of the natural world than in other areas. Iceland is one of the best spots on Earth for such an observation. Let us start with continental drift: In Iceland the ridge between the NorthAmerican and the Eurasian plates – which runs the whole length of the Atlantic Ocean – appears as a land surface. In other words, Iceland is the Mid-Atlantic Ridge come to the surface. The two geological plates are in the process of drifting apart at the rate of about two centimetres per year. That does not seem very much, but in a century it amounts to two metres and thus twenty metres in one thousand years.

Towards the edges of these diverging tectonic plates the crust is very thin. This means that the weight of recent lava in south-western Iceland is pressing the crust to subside about fifteen centimetres in a century. Toward the south-east corner of Iceland, on the other hand, the Vatnajökull Glacier is receding and losing weight due to global warming. This reduced weight will mean a rise of south-east Iceland of about two metres in this century.

In many other places of the world the crust of the globe is also subsiding or rising. In south-eastern England there is now a subsidence because of less precipitation that means less volume of sedi-

ment geological layers. This may have even more consequences than the expected changes in Iceland, since this corner of England, where London is located, is rather flat. The Londoners have already built flood gates to prevent sea floods from entering the Thames and rushing up into London. This fight against tidal floods will be harder in the future as in addition to the subsidence, the sea level will rise – perhaps as much as ninety centimetres in this century. Consequentally floods in the lower Thames will be a huge problem in London in the future, causing grave problems in the sewage and the metro systems, even though people will probably not have to be permanently relocated, as in some other countries, like Bangladesh.

A good way of illustrating the dynamism of Earth is to realize that maps are only a "snapshot" of the status of things. Because we look too uncritically at such static snapshots reality, most of us believe that the outlines of coasts, glaciers and vegetative growth are static, whereas they are in fact constantly changing. The weather machine of the world is now beginning to work at higher speeds. This machine drives currents, ocean flooding and erosion; forces that constantly batter the coasts of oceans, lakes and rivers. Coastal erosion is already of great concern, not least since coasts today are very popular with vacationers. The large new tourist coastal resorts are therefore, in many places, at risk.

In very many cases the necessary precaution of

More food barriers like this one in the Thames, need to be built in rivers as the sea–level rises.

At all times gobal warming (and cooling) have meant that traditional environments, landscapes and societies have changed. The warming of Greenland will mean that the traditional hunters' society will change to a modern society, as we know it in the developed world.

keeping a safe distance from the calm-appearing coastal waters is not heeded. We saw the consequences in the Indian Ocean at Christmas in 2004, as large settlements of tourist resorts were hit by an earthquake-induced tidal wave that killed about 200,000 people. The rise in sea level will mean that such floods, caused, for example, by the increased ferocity of winds and the increased number and size of hurricanes will subject certain coastal areas to catastrophic danger. Here again, we need to accept as a basic feature that the dynamism of these natural forces is overpowering, and make coastal plans accordingly. This requires a *total rethinking of the coastal planning* and the *coastal defences* in the world in the next few decades.

Becaus we have started to be aware of myny problems that come with global warming the idea is gaining ground that our civilization is headed for a total collapse due to our wrong and excessive ways. In 2005 the American author Jared Diamond published his book *Collapse*, which has caught a great deal of attention. This book

describes, with numerous examples, how past civilizations have collapsed. The author chooses to interpret the collapses mostly as something that these cultures could have avoided, had they behaved properly in terms of coexistence with Nature, hence, its subtitle: *How Societies Choose to Fail or Succeed.* This implies that today it is now in our own hands to decide if our world is going to collapse, and that it is almost certain that this is going to happen if we do not follow certain rules of conduct which scientists and some politicians have outlined for us.

This present book tries *to present a contrary vision of global warming* and its possible consequences; *a positive vision,* supported largely by the fact that global warming has happened several times before, and taking the view that we can act in time to change our behaviours and our use of space to compensate for and adjust to the changes. Rather than zeroing in on what may be looked at as a collapse of certain lifestyles and cultures, it is also possible *to interpret these occurrences as simply the world's natural way* – resulting from climatic fluctu-

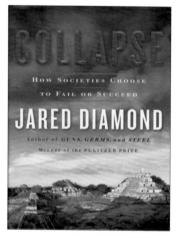

This book explains how some societies have collapsed because of climate change.

An old myth tells about the Phoenix that periodically burns to ashes so it can be recreated. Great disasters offer, in a similar way, an opportunity to start a new structure from the ground up.

Let's Embrace Change!

Not to fight change
Is a lesson to be learned.
The new way of life;
Is to embrace change!

ations that we can do very little about – and instead of trying to fight the changes, we should rather try to learn to live with them and to adapt to them.

It is certainly true that the Nordic settlement in Greenland collapsed and disappeared during the Middle Ages, as Diamond points out, most likely because of global cooling and the resulting lowered grassland yield. Also the increased extent of pack ice and the difficulty in keeping shipping lanes open to maintain ties with the Nordic world were very desissive.

Following this disscision it is quite logial to ask: Were the changes that came with global cooling in Greenland not just the natural way things happen, and should happen? And, if we want to interpret this "collapse" in a positive way, we could say that the Inuit nation of Greenland, in the process, reclaimed their land, something that probably was positive, seen from the Inuit point of view.

There have certainly been disasters that have hit the world that we would rather not have had to deal with, like the *Black Death* in the mid 14th century killing between a third and two-thirds of Europe's population, or the two *World Wars* that killed about 60 million people. In these cases, again, we can interpret them as disasters, and also, surpicingly, as something that in addition had a positive side to it. The epidemics can even be interpreted as Nature's way of stopping overpopulation, which in many cases, is the main reason

Wars, plagues, social upheaval and global warming mean terrible sufferings, but they are also an opportunity for a new beginning.

for most environmental disasters.

When it comes to the two World Wars we know that – on the positive side – both wars were catalysts for great technological advances, advances made in a very short time span, advances that otherwise mighth have have taken centuries. Though usually not recognized, Word War I, for instance, was the impetus for important medical advances. Social structures were also broken up and, due to the chaotic situation in many countries, the monarchies of for instance Russia and Germany were overthrown, paving the way for democracy or communism.

The lesson from this is that change most often brings hardship and challenges that people are afraid of and suffer from, and may die of, but that *change also means a force for restructuring* and *adjusting to new times,* which might otherwise have been e ven more painful and in addition have been almost impossible to accomplish.

Therefore, our attitude towards change should not be solely negative and in the case of global warming – as it seems certain that it will happen – it might be wise to meet it, and the numerous changes and challenges that come with it, with a positive attitude, and even with the cry, *"Let's embrace change!"* This "new philosophy" would mean that *we should go with the changes,* as our forefathers did. This requires an attitude of humbleness, an attitude that parts with today's haughty attitude towards Nature and life.

We need to accept the supremacy of natural forces and prepare ourselves to make it our primary responsibility to accept and study all the uncertainties and changes that are coming to us. We need to be ever humble and we need to rid ourselves of all haughty attitudes of religion, culture and race. This is absolutely necessary because only together, as one unified human race, can we meet this huge challenge of securing lives on Earth in the face of changes that global warming is going to bring upon us.

2 Changes in Weather Patterns is Nothing New

Human memory in terms of weather is very short. Certainly, the world has been experiencing extremes in weather of late, but such periods of extremes are, however, nothing new. There have always been occasional periods of warm and cold, wet or dry – periods that have had very severe impacts on nations that mostly live off the fruits of the land.

Generally speaking, the global climate has been *getting warmer by some 0.8°C* in the last one hundred years or so. Most scientists think that the warming of the globe this time around is happening because of the extreme *anthropogenic release* of greenhouse gasses into the atmosphere. As the emissions are predicted to increase enormously in the future – if special measures are not taken – the world temperatures will continue to rise.

The main changes in weather patterns that come with global warming are increased precipitation, more strong winds and also more aberrations in climatic patterns in terms of time and geography.

In the discussion of the impacts of the weather extremes that the world has been experiencing, the media, and even scientists, seem to have largely forgotten that extremes in weather have been hitting regions and countries throughout all of Earth's history.

As a point in case the table to the right shows *casualties from extreme floods* in the last 75 years. It shows us that even the earthquake-induced tsunami flood in the *Indian Ocean* in 2004, which claimed about 300,000 lives, is only one in a string of catastrophic floods that have claimed up to about 3 million lives. Such extreme weather events often severely impact vegetation, construction and transportation, as shown in the matrixes on pages 22-3 and pages 26-7.

Understandably, with the prospect of increased changes in weather, scientists have started to study more thoroughly what the various types of weather changes have meant for earlier civilizations. One of the books examining past weather-caused catastrophes is *Collapse* by Jared Diamond. That book, as the title indicates, focuses on cultures that have collapsed, mostly because of environmental problems. By studying Diamond's examples of past environmental problems, some useful knowledge can be gathered on how to meet the climatic changes of today and of the future.

As the author points out, climate change was even more of a problem for early societies because of less possibility to record the earlier experiences and to learn from them. In many cases periods of, for instance, wet weather exceeded the span of human memory. Early Greek philosophers believed in the cyclical nature of history and some books of old wisdom, like *the Old Testament,* tries to remind readers that *everything goes in cycles* and that the good years should be used to prepare for the meagre years. Even with a well recorded knowledge of cycles, humans have had a very hard time thinking about the consequences of weather cycles and do not heed the advice that every action should be planned with such cycles in mind.

In the past, many societies lived in isolation, or surrounded by enemies, so even a hazard of very local influence could mean the collapse of a country, whereby the enemy frequently claimed the devastated territory.

One of the good things about the modern *globalization* of trade and relief efforts is that local calamities do not hit people as hard as earlier and succour may come from people and organizations who are far removed from the site of the calamity.

In his book J. Diamond lists the features that are *most damaging to the environment* and identifies the following eight categories: *1. Deforestation and habitat destruction; 2. Soil problems* (erosion, salinization and loss of soil fertility); *3. Water management problems; 4. Overhunting; 5. Overfishing; 6. Effects of introduced species or native species; 7. Human population growth;* and *8. Increased per capita impact of people.*

Most humans are little aware of that everything in nature goes in cycles: Good years are followed by meagre years.
 Many old books of wisdom try to convey this truth.

Indian Ocean	*2004*	*300,000*
Central America	*1998*	*10,000*
(2 million left homeless)		
Bangladesh	*1991*	*130,000*
Tailand	*1983*	*10,000*
(100,000 contraced deseases)		
Bangladesh	*1970*	*300,000*
China	*1959*	*2,000,000*
China	*1938-39*	*1,000,000*
China	*1931*	*3,000,000*
(incl. resulting starvation)		

Extreme floods are more common than we realize, and the number of casulties higher.

1. *Human-caused climate change*

2. *Build-up of toxic chemicals in environment*

3. *Energy shortages*

4. *Full human utilization of Earth's photosynthetic capacity*

Four environmental problems that in Jared Diamond's view, will be a cause for tremendous societal problems in the future.

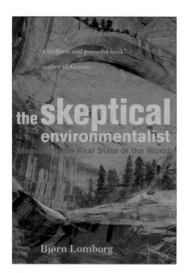

This book corrects many statistical data that is used in the environmental discussion today.

Most of these problems and processes in nature are linked to possibilities provided to yield agriculture production.

Today agricultural production or fishing can, on account of highly advanced ways of dealing with environmental changes such as genetic modification of crops, be almost completely *controlled by man-made changes* in environmental conditions. Therefore, *these early types of environmental damage* will probable only very locally mean a collapse if the damage is very excessive, as we certainly have seen happening in some places on Earth during the last decade. A total collapse of a country or culture, as in *Somalia* and *Rwanda* recently, will probably be only rare and exceptional cases.

In his book Diamond discusses the risk of comparable collapses today, which he says are of increasing concern to many, especially in very poor, third world countries. Diamond identifies four environmental problems of today that will probably be the cause of tremendous societal problems in the future. These are: *1. Human-caused climate change; 2. Build-up of toxic chemicals in the environment; 3. Energy shortages;* and *4. Full human utilization of Earth's photosynthetic capacity.*

In our study of such possible causes of the global problems in the future, one should keep in mind that *there are also other global problems already here or on the horizon* that could mean huge problems on a regional, or even global, scale. The table on page 31 showed us a list of these, including, for instance, communicable diseases like HIV/AIDS – governance and corruption – and malnutrition and hunger, not so much because of lack of food but because of poverty.

In all periods in history *there have always been fears about looming disasters.* Fortunately, people have in most cases been able to escape the disasters, often because the fear induced humans to take the necessary precautions and disaster was averted. People who tend to be afraid or insecure about the future may often have an inbred psychological tendency to view things bleakly and negatively, and we often talk about these people as *Doom-Sayers.*

It is very easy to put together long lists of predictions that some people "foresaw" as a very serious threat to humankind, threats that, however, never materialized. Why they have not materialized, might in some cases be because of the fear-induced proactive measures that were taken. These concerned citizens – who are today mostly called environmentalists – certainly have a very valid function in society, not least in our crowded world.

In some cases – as with the possible prospects of serious impacts of global warming – the discussion by itself is a way of preparing us for the consequences in such a way that the possible hazards can to a large degree, in many regions, be eliminated, as earlier sections of this book have pointed out.

In many cases today's environmental discussion is conducted in the media by rather naïve but influential people like actors and artists, and also by politicians who have power but not necessarily a background to understand the comlexity of the environmental issues. Many scientists are also so specialized that they only have a tunnel vision of the world, which means that they do not have the overview of history to understand that the climatic processes that Earth is experiencing today are not much different from many other periods in Earth's history.

Many scientists and artists also have very little understanding of how capable *science* and *management* are in coping with problems that certainly, at earlier periods in time, would have meant catastrophic disasters. Many such problems can today be dealt with by thorough planning and management, as well as by technological innovations.

The climate and the natural environment are very complicated fields in terms of scientific studies. To make matters worse, statistical material that covers large time-spans hardly exists. Therefore many scientists have had to rely on assumptions and unreliable statistical data to make projections about what will happen in the future.

The Danish statistician *Björn Lomborg* has written a book on faults in the theoretical foundations of the environmental discussion of today. The title of his book is *The Sceptical Environmentalist – Measuring the Real State of the World.* What Lomborg is mainly questioning are the so-called *State of the World Reports* that are periodically published by *The Worldwatch Institute*. Lomborg's work has ignited the fury of environmentalists as he lays bare the fact that many statistical interpretations in so-called scientific reports are false and not actually the work of professional statisticians.

The publisher of the book introduces it in the following way: *The Sceptical environmentalist challenges widely held beliefs that the environmental situation is getting worse and worse. The author, himself a former member of Greenpeace, is critical of the way in which many environmental organizations make selective and misleading use of scientific evidence. Using the best available statistical information from internationally recognised research institutions, Björn Lomborg systematically examines a range of natural and environmental problems that feature prominently in the headline news across the world.*

There is no doubt that Björn Lomborg has done very useful work in terms of getting more rigour injected into this environmental discussion, but he takes it a step further and concludes that there are more reasons for optimism than pessimism. There he probably oversteps his abilities; it is not possible for anyone, statistician or others, to say that the very grave concerns that many people have today for the future are not a cause to be pessimistic. On the other hand, Lomborg's optimistic attitude is by itself a very good way of dealing with some problems, even if the degree of optimism may not be totally warranted.

It is seminal to review Lomborg's main findings and his conclusions will be given: In the conclusions of *Part II: Human-Welfare*, Lomborg stresses that we today are experiencing unprecedented human wellness: *"The number of hours we work has been halved during the last 120 years…. The murder rate has fallen considerably….On the average, we have become much better educated, and the developing world is catching up…."*

However, Lomborg does not close his eyes to the problems, *"Africa stands out as a prime problem area, where African people have experienced much less growth over the last century than people in most other countries, an epidemic has engulfed parts of southeast Africa, and because of war and ethnic and political division the outlook is not rosy. But even Africa is still better off than it was at the beginning of the 20th century…."*

Part III of Lomborg's book answers the question: *Can human prosperity continue?* Some of Lomborg's main conclusions are: *"We are not overexploiting our renewable resources. Worldwatch Institute tells us that food scarcity is likely to be the first indication of economic and environmental breakdown.…[but] food will in all likelihood continue to become cheaper and more available and we will be able to feed still more people."* On the negative side he says: *"it is possible that we pollute so much that we are in fact undercutting our life."*

As the book comes to the study of tomorrow's problems in *Part V*, global warming features prominently. Lomborg quotes the IPCC 2001 report's summary for policymakers: *"Published estimates indicate that increases in global mean temperature would produce net economic losses in many developing countries…."* On this Lomborg says: *"This sends us two messages, first, global warming will be costly…, second, developing countries will be hit much harder by global warming…."*

We can summarize: Increased production of greenhouse grasses, in all probability, is responsible for the increase in global temperature. The problem is thus mostly caused by northern and western countries that have built up their wealth by unlimited use of cheap fossil fuel for their industries – whereas, on the other hand, the developing countries of the south will be suffering most from the resulting warming of the climate.

It should therefore, ethically, be of highest priority for the industrial north to help the southern, poor regions of the world to become industrialized and to play a larger, more important role in world markets so that they may have some leverage in terms of income to deal with the environmental problems that will hit them more than most other regions of the world.

Once again, the author of the present book wants to stress that we can only understand and effectively respond to global warming by developing and utilizing a *global overview*. This overview tells us that, at the same time that there are areas, mostly in the south that will suffer from global warming, there are other areas towards the poles that will gain greatly. *The balance* so attained is hardly entirely negative but will require considerable readjustment in our thinking and planning, in our *recognition and utilization of opportunities.*

"We are not overexploiting our renewable resources. Worldwatch Institute tells us that food-scarcity is likely to be the first indication of economic and environmental breakdown.…[but] food will in all likelihood continue to become cheaper and more available and we will be able to feed still more people."

On the negative side Lomborg says: *"… it is possible that we pollute so much that we are in fact undercutting our life."*

Some of Lomborg's main conclusions in his book "The Sceptical Environmentalist."

The hot and arid areas at Earth's centre will be substituted by areas that are now too cold, but will become fit for inhabitation with global warming.

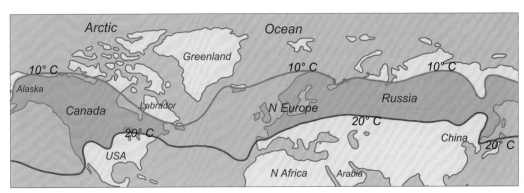

This picture shows how low the average summer temperatures are in the old industrial countries of the North. This means that they would be better inhabitable with a little warming, but their bio-zones, of course, would have to adjust to the higher temperature.

Let us adopt *a balanced and realistic attitude*. We can help ourselves in this task by realizing that there have always been global changes in weather patterns, even more extreme changes than today. Historically, humans and nature have responded to this in the very simple and realistic way of *"going with the changes"*, which means; to move to areas that are better suited for habitation, be it warmer areas, wetter areas, calmer areas, or whatever the weather patterns are causing organisms and humans problems in a given region.

If we study the impacts of global warming, it is very interesting to note that not much catastrophic is going to happen in northern industrial countries with the warming of the climate. This part of the world will be able to adapt quite well, in spite of problems regarding certain aspects. A central conclusion is that the developed world is, and will be, ethically obligated because it has mostly caused the problem of global warming, to introduce drastic measures to help the developing world with not only climate induced problems, but also with other tasks, such improving welfare and wealth, because if such primary conditions are not improved these nations will not have the strength to deal with the environmental problems.

This needed action in development aid, however, is not only based on moral responsibility but also on the realization that the world has become so globally interdependent that problems occurring in, say Africa, will almost certainly have severe consequences for most other regions of the world.

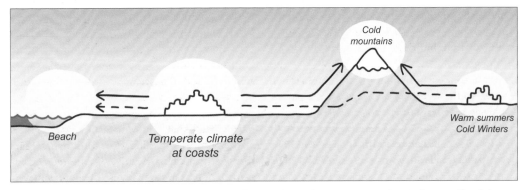

Rich people have huts at beaches and in mountains where they can go to during hot spells. Other people can go there on their holidays for cooling.

3 How Global Structures Change

This middle part of the book is called *Patterns,* this present chapter *How Patterns Change* and this present section *How Global Structures Change.* These chapters involve an unusual use of the word *pattern* and also the word *change,* so a clarification is needed.

First of all, a short introduction: As explained in the first section of this chapter, the emerging worldview is characterized by active time. In other words the meanings of the words and concepts linked to time – such as change, development and pattern – are becoming ever more important.

The word pattern is a part of the context created by active time and is used to signify how something happens over time. Dictionaries also give a meaning that signifies a *static pattern* as, for instance, the *pattern of a settlement. Dynamic patterns* – like the ones dealt with in this book – are not as easy to describe with pictures as the static ones; one has to resort to presenting them as a series of pictures, or arrows that show in which direction things are developing.

An easy illustration of a dynamic pattern is how settlements move towards cooler areas in times of global warming. In the case of such large-scale movements we talk about a *megapattern.* The term megapattern is also meant to signify overall spatial trends where local deviations are omitted.

The word *trend,* like the word pattern, also has its foundation in the *active passing of time.* In 1982 John Nesbitt published his book *Megatrends – Ten Directions Transforming our Lives.* The directions, or trends, that Nesbitt described are ways that societies in general are evolving, for example, from an industrial to an informational society and from a national to the world economy.

The present book does not work with trends that originate in the social realm but rather with the *spatial tends* in nature and human settlements. Such spatial trends are in most cases mobilized by changes in the climate of the globe. Because of this spatial nature of the word pattern, it fits well for signifying spatial trends. The phenomenon of

megapattern signifies large-scale spatial development and is the core of what is attempted in this book; in fact, an alternate title of the book could be *Megapatterns of How the World will Evolve.*

As we have now gained a somewhat clearer understanding of the core terms of this book – *pattern and megapattern* – we can proceed to another important term in the book – *structure,* a word used in the title of this section. In planning we know the word structure in combinations, like *settlement structure* and *utility structure,* sometimes called *infrastructures* because they are mostly invisible.

On a global level the best known *global structures* are the *transportation structures* on land, on the sea and in the air. These large-scale structures do not change much in a time scale of only decades, but if we look at developments over the last hundred years, we see many changes.

Looking to the next one hundred years – this century – we can be certain that the transportation structures of the world will change significantly, not least because of global warming. The third and last part of this book – *The Future* – will make an attempt to predict what the changes will be; what for instance, the global-settlement- and transportation patterns will look like at the end of this century.

The task of the present section is to describe how, in general, *global structures* have evolved and changed. In order to be able to understand how global structures change, it is necessary to get some understanding of what features have had an influence on how they have changed and evolved, and then add thoughts on some of the basic premises of the evolution of the past and how the structures may change or evolve in the future.

The traditional way of studying *settlement patterns* is by means of static geometric patterns. The dynamism involved is most often largely disregarded. This is mostly because of an underdeveloped understanding of the fourth dimension-time in today's society and in today's science.

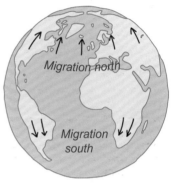

Megapattern forged by global warming: Towards the poles.

Megapattern forged by global warming: To coasts in warm countries.

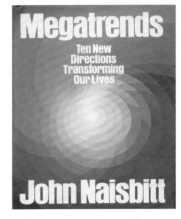

A book that prioneered the study of overall trends on a global scale: megatrends.

By looking at large time scales in the history of the Earth, we see how the world is constantly changing, both the natural world and the man-made world. Earlier, people even used to think that nature was static. Global warming is teaching us that our world is not static but rather very *dynamic.*

Let us now review the development of megapatterns in the history of the world. Historians often place the beginning of "modern" man's history at about 10,000 years ago *in the Near East,* as the globe started to get warmer after the last ice age. The first stages in this history were mostly characterized by the ways in which humankind fended for their living, first as hunters roaming the woods and plains and then, somewhat later, in more organized groups that we categorize as nomads.

What made this shift possible was that humans, notably in the Near East, had succeeded in domesticating animals useful for their livelihood. At first humans roamed around with them in search of water and the best areas to feed them, same as Samish (Lapps) in Scandinavia still do.

One of the most important stages in the development of human culture was fixed settlements that accompanied the development of agriculture, leading to the rise of farming communities, which in turn led to the development of trading centres that also became governmental and economic centres. This has happened at very different periods of time in the different regions of the world.

Farming crops and tending livestock required daily care and remaining in areas where the water supply could be counted on. The best areas for agriculture and husbandry, was where the climate was benign and large fixed settlements could develop, mostly by the large rivers that empty into the Persian Gulf, the rivers of India and China, the Nile in Egypt. The cultures of these areas reached such heights that they still today inspire admiration.

An important part of being able to form fixed settlements was to have the means to transport goods to and from the settlement – at first mainly human portage or on rafts on slow and calm rivers. Eventually domesticated horses, oxen and camels also served transportation purposes – which also happened in different periods in different regions.

The biggest step in the development towards today's settlement structures was the development of ocean-going ships. People who could build such ships were at a huge advantage in terms of transporting goods to and from even very distant places. Actually, Viking outreach from the ninth and tenth centuries on was possible in large part because at the time they built the only ships possible for regular ocean voyages, albeit only in summer. With improved ship design the balance of power shifted to Spain and other countries as they explored, settled and traded. With the advent of even more seaworthy and faster ships, from clipper ships to steamships, trans-ocean trade expanded exponentially.

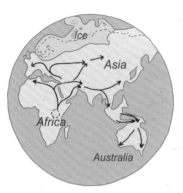

How humankind spread out of Africa 150,000 to 18,000 years ago, e.g. before the last Ice Age.

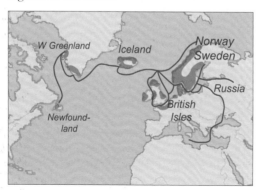

Technical ability in transportation has always been a strong factor in historical development: Here the main travel routes of the Vikings.

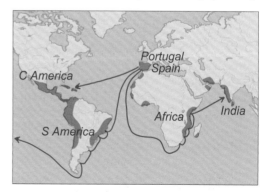

Their skill as seafarers enabled Europeans to start colonializing the world in the 15th century. The Spaniards and Portugese being the first.

Conquering and colonization, from the sixteenth century on, also followed in the wake of improved ships. But whether settlements had been established by foreign colonizers or earlier by the local inhabitants, understandably, the ones that thrived were the coastal settlements that had good, natural harbours.

One of the prerequisites for creating and sustaining large societies was the possibility of organizing large armies. At first armies mostly travelled overland in their ruler's quest for valuable land and commodities, but with the advent of *ocean-going ships* they could go farther, making sudden raids on coastal settlements that were at first even less protected from an invasion from the sea than from the land.

Even small tribes were capable of going long distances on the ocean, gathering enormous booty, if they had good weapons and good ships. A prime example was the case of the Vikings who principally raided Northern Europe, mostly the British Isles, but also the Atlantic coast of France, into the 9th century, sometimes going up rivers like the Seine in order to reach large settlements like Paris. The Vikings also went via the large rivers of Eastern Europe all the way down to the Black Sea, where they carried out business and raided settlements, all the way to Constantinople (Istanbul) and well into the Mediterranean and to the coast of Africa.

As political and social organization developed further – mostly with increased transportation possibilities – quite large national states were formed. Eventually, the power of some of them spread further and further, like that of Rome, which became such a far-reaching power that it colonized or captured most of the Mediterranean space and Western Europe. Centuries later the North European tribes started to form societies, led by successful warlords, and the tables were turned; the tribes that had previously been held at bay now swarmed over the former borders of the Roman Empire, invaded and sacked Rome in 410 A.D, and succeeded in destroying the Roman Empire.

The Middle Ages witnessed the start of the development of cities and of western states, which, through the impetus of nineteenth century nationalism, became nation states as we know them today. These states eventually grew in wealth so that they were able to build large ships and go still farther in search of valuable land. The most famous of the exploration journeys was Columbus's search for Japan in 1492, which ended in his bumping into the Caribbean Islands and America 500 years after the Norse had already discovered North America. Columbus's discovery started the settlement of the white man in the new world of the Americas.

A little later or in 1498, the shipping route beyond the tip of South Africa was discovered and beyond South America in 1521, which opened up the colonization of Oceania, the Asian sub-continent and China, coming from two directions. In the 13th century, however, overland transportation between China and Europe, principally the *Silk Route,* had already come into use. Particularly a Venetian family, of whom Marco Polo became the most famous scion, recognized and exploited the potential of trade with China, as did others of other nationalities.

The European colonies in Asia and Africa did not become permanent white settlements and eventually, during the 20th century, various wars and local conflicts ended in forcing Europeans to give up their claims. Remaining as new settlements, new lands, for the white man were the Americas, New Zealand and Australia.

The European expansion also took place over land in the far North, into the heartland of Asia. It was mostly the Russians who advanced in this direction, making the huge terrains of Northern, Central and Eastern Asia part of their empire in the 18th and 19th centuries. This area, totalling about 15 million square kilometres, reaches from Europe to the Pacific Ocean and covers 11 time zones.

The theory on how settlements evolve and prosper includes several approaches. One of them is the *central place theory,* a theory that says that a place or an area that is centrally located in a country, or located at strategic places in a transportation system, like crossroads, are those which prosper. A further elaboration of this theory is the so called *Christaller model,* which explains how certain geometric patterns develop on land in

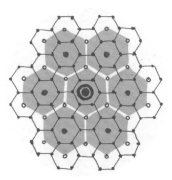

The Christaller model: *Explains how a hierarchy of centres follows levels of hexagonal patterns.*

The Silk Route: *The linking of Europe and China was a large event in the world's history.*

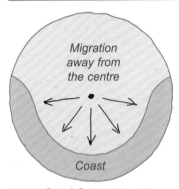

Centrifugal force: *Away from an interior towards coasts. A force in how settlements develop.*

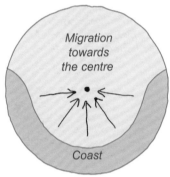

Centripedal force: *Towards the interior ... away from coasts. Mobilized by interior development.*

Capital at the centre: *The best location to serve all equally well. Here Madrid as centre of Spain.*

terms of settlement structures, forming a crystallic hierarchy of regional centres, local centres, etc.

The main impetus for the development of towns was that an authority of some kind settled there, for example a secular or religious power, frequently protected by walls, a format that became very common in the 12th century in Europe. In the agricultural areas of Europe various types of processing industries eventually developed for food production and commerce, and settlements that specialized in these methods were formed in the centuries to come.

Large-scale *fishing* also eventually led to the forming of towns, for instance in Northern Europe in the Middle Ages, but a rather advanced commercial system, like the Hansa system, was needed because the fish had to be exported and traded for other goods.

The development of new types of industry, in most cases, led to the formation of urban cores. The first step in this historic development was *primary production,* including that of salt, fish, metals and coal. A second stage was industries that worked with and *processed the primary goods.* The third step was the development of *service industries,* including services for other basic industries like blacksmiths and mechanics, and services that facilitated enlargement of settlements, like food production and construction of buildings and infrastructure.

In some of the more advanced countries special *centres of culture and learning* developed. The most recent of the basic types of urban settlement – a type of town that is still under development – are *leisure-based towns.*

As we have now seen, there have been certain localities of natural resources and human activity that *attracted* people, but there were also other features that *repelled,* as for instance hazardous areas and areas of social unrest and overbearing warlords.

The settlement of an area therefore actually develops within a *field of forces.* Basically, there are two types of fundamental forces: a *centripetal force* that pulls people into an interior area, away from coasts, and secondly, a *centrifugal force* that sends people out of an area toward the coasts. Transportation on rivers and land-based trans-

portation, contribute to a centripetal force, whereas shipping and fishing are centrifugal forces which pull people from the interior of a country to the seaside.

In the early 19th century, at the beginning of the industrial age, most industries were dependent on good solutions for the very heavy transportation of iron and coal. At that time ships and barges were the only economical way of transporting heavy cargos over long distances. Therefore most of the first modern-day industrial towns were placed along coasts and navigable rivers. In this era seaside and riverside towns and ports prospered, leading to the building of even better ships – steamships, that opened up the way to sail between continents in all seasons. This meant a huge spread in *globalization* of trade, a development that led to even further prosperity for coastal towns.

Because of the migration to coastal settlements, the population of interior areas was reduced. What started to reverse this settlement process of pulling people to the coasts was the strengthening of interior transportation with canals, locks and larger motorized barges on rivers and the development of railway lines.

Industrialization came with a time lag to some of the less developed countries. These countries eventually began to go through the same pattern, where the first modern towns, such as Lagos and Mombasa in Africa, started to develop along the coasts. The development of the interior with river, rail and road transportation came later to some areas, and to others is still developing.

The development of the interiors has now reached such a degree in Western Europe that today most coastal towns have declined considerably because, with a few exceptions like Rotterdam and Hamburg, these societies are no longer based primarily on heavy industry and/or heavy transportation. Transportation in these modern societies is principally that of light goods and people, via roads and trains and – since the early 20th century – airplanes.

Airplanes have been a strong force in developing some centrally located areas deep within continents, like the city of Brasilia that was decided on, planned and established because of its central

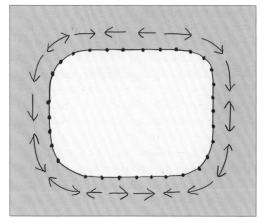

1) Transport and resources along the coast:
The coast a linear centre without point centres.

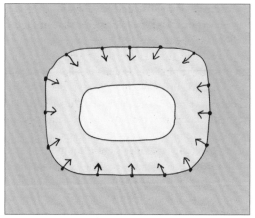

2) Opening of areas next to coasts: centripedal
force. A ribbon of habitation is created.

*Six phases in the spatial
development of a region.*

*Western Europe developed
according to this scheme and
the regions at the Arctic Ocean
will also go through these six
phases of spatial development.*

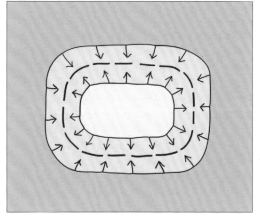

3) A new linear centre develops in the middle
of the ribbon: The coast loses importance.

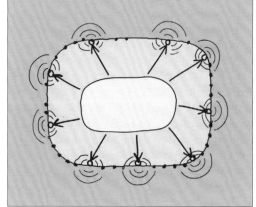

4) World shipping and fishing: Large harbours
get developed at coasts.

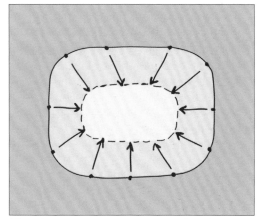

5) Less shipping and heavy sea transportation:
Interior development. A pull to the interior.

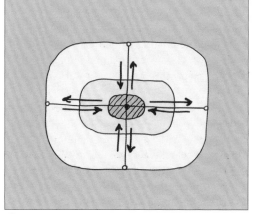

6) Crossroads and settlements at the geographical centre: An interior centre develops.

Declining pack ice will open new major shipping routes between the Atlantic and the North Pacific spaces.

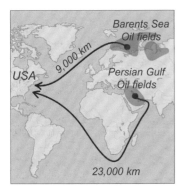

The shift in oil exportation from the Persian Gulf to the Barents Sea will have huge consequences.

location as the capital of Brazil. Brasilia is therefore called the *airborne city* – though, in fact, it has not been as successful as hoped as a centre for Brazil. Earlier, similar central locations were chosen for the capitals of countries, like Madrid in Spain and Mexico City in Mexico – because that location minimized and equalized the distances in the countries they serve.

A comparable *sequence in development* of settlements is very likely going to happen in the Arctic areas in the late decades of this century. These areas will at first mostly function as producers of heavy goods, like oil, minerals and timber. It will therefore be port cities that will primarily develop in these areas first, mostly at estuaries because large rivers that flow into the ocean are very convenient for heavy transportation once they are ice free.

Further inland transportation and interior settlement will, on the other hand, be very slow in developing.

The primary industries that will develop in the Arctic will probably not entice very many people, beyond what is necessary for their operation, since the Arctic will for a long time continue to be cool, and the long dark winters are not very attractive for all-year human habitation. In many cases people will only go there seasonally – as to the oil floating rigs today – and work there for a limited time, preferably during the bright summer months; a season without nights – and the leisure in the more benign areas of the globe to enjoy the fruits of their hard labour.

In the history of the settlement of the world we can easily observe how much influence transportation routes have had on the status and prosperity of certain areas. One of the most famous places, strategically, has long been the Strait of Gibraltar, its importance going back to Phoenician times. After the Suez and Panama Canals had been built, these areas acquired strategic importance. Egypt gained political power and Panama, largely a US creation from a part of Colombia, has become something of a bone of contention between the native Panamanians and their government and US demands to ensure protection of the passage of the US naval fleet more rapidly between the country's Atlantic and Pacific

coasts. However, all ships passing through canals are limited in size by the width of the canals and the size of the locks. In the fall of 2006 the people of Panama desided to extend the capabillity of the canal system to allow 12,000 container unit ships from today's 4,000. This will mean that the importance of the Arctic shipping routes for this ship size category, for certain markets, will decrease.

The shipping lanes through the Arctic Ocean – connecting the North Atlantic and the North Pacific – will, however be of enormous importance as these lanes will be much shorter shipping routes for instance between South Asia and the North-Eastern USA and Northern Europe. This will mean that shipping through the Arctic Ocean, with the reduction in the coverage of the pack ice, will not only increase, but that tankers cargo ships can be still larger than they are today. 500,000 tons tankers and 15,000 to 20,000 container units ships are already on the drawing boards. Plans are to deepen the Suez Canal to 21 m by the year 2010, which would allowe this size of ships, but further enlargment is unlikely because of the limit in the Malacca Strait.

The development of the new Arctic transportation routes will mean quite a fundamental change in the transportation and settlement structures of some areas of the world. As an example, the importance of the sailing route through the Mediterranean and Suez Canal will be reduced, as will the sailing route into the Pacific through the Panama Canal. On the other hand, the sailing routes north of Canada and along Iceland and Norway into the Arctic Ocean and through the Bering Strait to the North Pacific will be strengthened.

Many other factors like the development of certain other spatial systems of the globe will also greatly influence how the settlement patterns of the world will develop in the future as explained in diagrams on page 82-3. In the third and last section of this book, *The Future,* that starts on page 79, the elements and details of the new global structure that comes with the activation of the Arctic, will be described.

2 The Static World of Today

In early periods of history as people were living as nomads they moved around in response to changing conditions, just like birds and animals do. Their adjusting to climatic fluctuations was therefore just a part of the process of going with nature, both on a local and a global scale.

As *fixed settlements* started to develop in various regions of the world the adjusting to climatic fluctuations became harder. People of this static way of surviving had to live with the spells of warmth and cold, wetness or drought that occurred in the regions where they were, even though moving over rather short distances to a better area – like the nomads did – might have brought them out of the problem area. Sacred texts, like the Bible, include recommendations on security measures to help *remedy the faults of fixed settlements,* for example by storing grain to compensate for the droughts that were sure to come.

In very early times these *fixed ties to locality* were more difficult than today because in-house and microclimates could not be manipulated as they can be now. Today we can heat or cool our houses and today, in agriculture, we can create shelters, install irrigation, cover areas with plastic sheets, and develop and change the crops grown to maintain agricultural production in dry periods.

The growing *technological ability in mastering climat-ic deviations* is very reassuring as we, with global warming, can expect more climatic extremes. One can even envision that a thriving settlement could exist under glass in a desert, as was tested in Arizona with the building of *Biosphere 2* in 1989. A model for this idea of a totally controllable environment is the so-called glass bubble experiment, where plants maintain the balance of CO_2 and oxygen and where the same water is endlessly recycled. The same principle is used *in space stations* where human waste creates new food and purified urine is recycled as drinking water.

The problem with such technological solutions is that – even though possible – they are out of reach for poorer people, and are also in some cases energy consuming, which remains a huge problem as long as inexpensive ways to produce clean energy have not been invented.

In addition to these technological ways of dealing with climatic fluctuations, over time humans have *created many ways of adjusting to climatic conditions* in an area.

In *northern countries* settlements are placed in areas inclining to the south and west to maximize exposure to the warming rays of the sun and trees are used to create shelter and reduce the effect of cold winds, like the cooling of buildings. In *very warm countries,* on the other hand, settlements are

Both the natural and human worlds are very dynamic, which is in conflict with today's static schemes and modes of thinking.

Biosphere 2 was built in the Arizona desert to test how a closed ecosystem and closed human community would thrive in isolation.

Many animals and birds migrate with seasonal changes. Still today some Samis of Scandinavia migrate with their reindeer herds.

In northern areas settlements are placed in sunny spots but in shadows in very warm areas.

People migrate as birds do. On holiday northerners go to warm areas and southerners to cool areas, a choice which is increasing with global warming.

Sometimes the problems associated with global warming are exaggerated – as here in the picture: The flooding of the coastline of Manhattan by a 5 metre rise in the water level.

The faults of this presentation: The rise in sea level in this century will "only" be 50–90cm – and also: it is, relatively speaking, not very costly to protect urban areas, as the Dutch have demonstrated.

located on the shadow side of hills, and trees cleared to open the area to a cooling breeze.

In spite of a mixture of non-technological and technological means to make settlements more comfortable, one should try to keep *the model* in mind that most organisms operate on; i.e. *to change locations and migrate in response to changing conditions.* Migrating birds do this and humans do this to a larger degree than we usually think about. Citizens in warm cities, for instance, go to cool coasts, cool mountains and, of late, to northern countries during hot spells, and inhabitants of cold countries fly to warm tourist resorts to enjoy a warm climate. But again, this is costly and energy consuming so we people on Earth have to be rich enough, and have enough cheap energy, to make this a generally applicable option of responding to climatic variations on a global scale.

The flooding of *New Orleans* in 2005 starkly revealed the problems involved in the application of this method of relocation; those who did not have a car to move out of the city, or money for lodging in other areas, *could not move because of their poverty.* Those who for example had mobile homes and caravans were well off.

The world has a *long history of people having to flee out of problem zones.* Often the cause has been war-

fare or some environmental hazard, like flooding or drought. Fortunately, *international relief methods and organizations* have been developing response measures, because large hazards often overwhelm the stricken country. Even the very rich and large USA welcomed international aid during the New Orleans flood.

The number of *environmental hazards* will increase greatly in the future due to increased environmental extremes and bad environmental management.

The first method or strategy that should be employed to reduce these increased risks is to *define the hazard zones* and enlarge the zones that have already been determined as such, and then *start to move people out of such areas.* Secondly, people need to adapt a *design and planning philosophy* that always seeks the most secure solution, a philosophy that counteracts the haughty tendency of today to challenge the forces of nature, and to court disaster in terms of plans and construction, rather than to develop sustainable solutions.

We have now for a while discussed environmental hazards on a local and regional scale that can mostly be dealt with within national borders. If global warming continues as projected, some regions of the world will *experience problems that can only be solved by extensive relocation* of large numbers of people.

The most graphic of the hazards is the *rise in sea level* that some predictions say can reach up to 90 centimetres in this century and, if the warming continues, could become many metres. What makes this hazard more manageable is that the rise in sea level will happen very slowly or at the rate of less than one centimetre per year.

Actually, adequate *coastal defences* can be built – especially by the richer countries such as the *Netherlands,* though for *Bangladesh,* which is also a very low country, the problem will be overwelming because the funds needed to build the defences will probably not be available. Another catastrophic environmental problem that will eventually engulf large spaces is the *spreading of deserts.*

In many cases it will not be possible to solve such huge environmental problems within nation-

Areas that would be flooded if the sea level rose 5 metres at Manhattan – that is if the shoreline were not protected.

al borders, neither spatially nor economically.

Fortunately, *large economic unions and unions of states* have been developing in the world, creating strong wholes that are able to deal with such problems. *Examples of such unions are* the USA, the EU and the Russian, Indian, Chinese and Australian Federations. Areas that are still mostly broken up into *rather weak national states* include Africa, Central America and a part of Southeast Asia. In such areas migration and mutual help are much harder to effect.

Even though the larger state unions are in many ways capable of dealing with such problems, their outer borders usually restrict the immigration of people from problem areas. For some unions exemptions are made for political refugees and comparable rules for *environmental refugees* are being discussed and formulated.

Today it does not sound very likely that these unions will open their borders to make migration from poor environmentally stricken countries any easier.

The natural direction of migration of the future problem areas of the Sahara rim is to Southern Europe and from the Middle East into Russia. This migration to Europe has already started, mostly because of poverty, as the present large number of attempts of refugees from Africa into Spain and Southern Europe attest.

The southern European states of Spain, Portugal, France, Italy and Greece already have a very hard time absorbing and integrating, and sometimes rejecting, this migration, and Russia already has large problems with its Muslim population and neighbours.

Some decades ago there was a huge demand in Western Europe for a workforce from less developed southern countries, but with the further development of automatic manufacturing processes and the developing third world economies the need for this workforce – and a workforce in general – is decreasing. The result is a huge problem of unemployment, not only of the less educated immigrants but also of the younger generations in these countries. This has already led to increased hostility between the foreigners and their hosts.

Every large and diverse country needs an enemy to strengthen its internal cohesion. In the case of the USA the enemy was Japan and Germany during World War II and one of their best friends, the then Soviet Union. After the war, Japan and Germany became best friends of the USA, granted most-favoured-nation trade treaties, and the earlier friend Russia, one of the worst enemies. Now Russia has joined the ranks of one of the best friends of the USA and Western Europe.

For some period there has been *a growing hostil-*

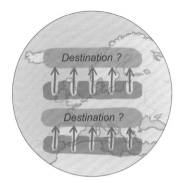

With global warming the climate zones move north. A natural pattern of the future migration of people would be that they would follow.

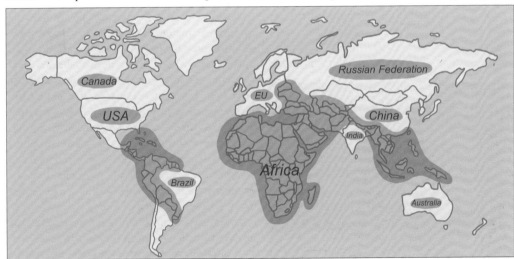

Large economic unions and unions of states have been developing in the world, creating strong wholes that are able to deal with large problems. Areas that are still mostly broken up into rather weak national states include Africa, part of South and Central America and a part of Central and Southeast Asia.

Who is friend and who is foe is always changing. Fewer conflicts is a must in the future.

Today the world is dependent on oil from the Persian Gulf. The USA thinks its presence in the Gulf is needed to safeguard the supplies.

ity between the USA and the Muslim world, among other reasons, due to American support of Israel. This hostility escalated to catastrophic dimensions with the attack of Muslim terrorists on the World Trade Center and the Pentagon on September 11, 2001. This made Americans, persuaded by President George Bush and his colleagues, willing to invade Iraq in 2003, whereas, however, the control of this oil-rich country and the Persian Gulf region was the main underlying reason. The danger of Iran producing a nuclear bomb is escalating US worries.

The hostility between the Christian-Western world and the Muslim world has reached a very serious state. Today these problems are driving oil prices up and political and trade cnnections have become less sure. As a result many Westerners have developed a growing antagonism to the Muslim world.

This will probably lead to braking the extension of the EU to the south, and possibly not even of the inclusion of *Turkey.* These religious-cultural conflicts are therefore slowing down the integration process that is so important to the southern countries, many of which will be hit hard by global warming. These countries need to build wealth and modernize with the help of their oil before it runs out in a few decades. Their adherence to what the West sees as static and outmoded religious doctrines, but by the West as detrimental and not understandable is seen by them as justice and strength. This clash in views and beliefs is a very serious obstacle to the needed integration of the cultures of the world.

It is best to reiterate: serious global changes in terms of cooling and warming have often happened in the history of the world before. It is helpful for us now, as we are facing a period of excessive warming, to look to past experiences with such climatic changes, be it because of warming, cooling, drought, excessive rain, or whatever the climatic factors may have been.

The proposed *strategy of migration* in response to global climatic changes may seem surprising. However, we need only to look back about a century and a half to find an example of large global scale migration because of climatic changes, in this case cooling of the climate. The main driving force of the migration that time was the last period of the so called "little ice-age", which occurred in Northern Europe in the late 19th century. This very cold period became the main reason for the settlement of many regions in North America. Overpopulation in the Nordic countries, freedom from rules that restricted movement of people, and most of all the advent of steamships, made this 19th century mega-migration between distant countries possible in only a few decades. In a few countries political problems were also a force that drove people to America.

Today, what seems most surprising to us is that people from all nations and regions were allowed almost unrestricted entry to the USA and Canada. It helped, of course, that these countries were large and sparsely populated and that the immigration from Europe to America followed "hori-

Jerusalem is a holy city for three religions: Jewry, Christianity and the Muslim faith.

Conflicts between Muslims and Christians have a long history, that makes things more complicated. The picture shows crusaders advancing.

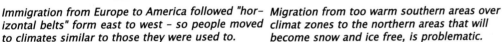

Immigration from Europe to America followed "horizontal belts" form east to west - so people moved to climates similar to those they were used to.

Migration from too warm southern areas over climat zones to the northern areas that will become snow and ice free, is problematic.

zontal belts" form east to west – with a southern tilt. The individual cultures settled in a latitude space similar to that which they were used to, but slightly warmer. *The inbred knowledge* of these cultures of how to build and cultivate were largely compatible in the new environments and were brought with them in their minds as blueprints. Minnesota, for instance, in the 19th century at first looked very much like Scandinavia. And the earlier 17th century settlers in New England built their trades and houses in a similar way as in their homelands in England.

In the extensive climate change that will hit the world in this century the situation will in some ways be similar to the situation of the late 19th century and in some ways different.

First of all, the climate is warming rather than cooling, and secondly: today there is no whole continent – stretching over several climate zones – available for the environmental refugees. Today the regions that will become habitable with warming are the sub-polar areas north and south of today's ribbon of habitation that goes around the globe.

The main problem that derives from this is that the environmental refugees of the next decades today mainly live in the south. Theoretically, there is an abundance of space free for them in Northern Canada, Scandinavia and Siberia, in the northern hemisphere – and in southern South America and southern Africa in the southern hemisphere. But people from the middle section of the globe are not likely to want to move to these "cold and distant areas." They would most likely prefer to move only a little north (or south) *following the spread of the biosphere* they know. This would mean that Arabs would like to move to southern Europe, Mexicans to the southern USA, etc.

The problem with this is that the climatic conditions in southern Europe and in the southern and southwestern areas of the USA are not so bad so that the people living there now are not likely to be willing to give up these areas for the immigrants, which might eventually lead to even stronger ethnic conflicts than these areas have recently experienced.

The migration process to the north described here has actually started, even though not yet really driven by global warming. This migration has already led to huge cultural clashes, most notably in France.

The regions that will be hit most by global warming are the areas that are already very warm, for example in the south of Europe and Russia. These are mostly Muslim areas - which have cultures, as history shows, which do not mix well with Christian cultures. The recent animosities,

Each year about 700,000 immigrants come to the US from Mexico in an illegal way.

In many Muslim countries there is still a strong connection between faith and politics, even if they are called secular.

which have culminated in terrorism, make it unlikely that Europeans and Russians will be willing to open their borders to a freer flow of Muslims into their countries or their regional unions. However, Turkey is most likely going to be a gateway for Muslim immigration to Europe, if it becomes a member of the EU in the future.

This now diverse cultural social dimension will be a decisive factor in how the migration of people from the too warm and dry southern areas to the colder, wetter and more northern (or more southern) areas can take place. It is therefore very unfortunate – given the more recent process of integrating and becoming one global community – that the world is now being disrupted by the increase of hostile tensions between cultures and religions, mostly between the Christian-Western world and the Muslim world, which also has an effect on other cultures.

Today, in a period of these cultural tensions that do not at the moment seem to have a good prognosis of resolution, we can, nevertheless, look for instance to the Americas to see examples of successful integration. The American countries – north, south and Caribbean – were settled by numerous cultures arriving from both the old world and from Asia and Africa, so we can see in many places that a mixing of cultures, religions and ideologies can really work. Of course, the integration process – for instance in the USA – took a long time. Some of the cultures and races mixed quite readily and formed the core of American society, whereas other races have remained more or less on the sidelines.

In the USA it was mostly the blacks who were *outside the immigration process,* since they were brought to America as slaves. Understandably it has therefore taken a long time to improve the status of the blacks. The 1960s was a decisive period for this integration in the USA with the leadership of strong and influential men such as John F. Kennedy, Martin Luther King and Lyndon B. Johnson. Today, blacks in the USA are visible in more parts of society than earlier, for instance in films, television, sports, and politics.

This integration process, in some parts of the world, as for instance in France and Germany, has been much slower and immigrants, blacks and

Many societies of the future will be multi-cultural, which will lessen tensions among nations.

Turks, for instance, are still today outsiders that hardly appear in the media. The old European nations need to adapt a more accepting policy to the new immigrants; to corner them in ghettos will only make matters worse.

Within *this area of integration* the European nations can learn a lot from the USA, in spite of the fact that they are in the habit of talking about the USA as treating its minority races badly. They do not see the plank in their own eye.

Of course, the culture of the future will be much different from the rather ethnically pure national states, as we still see them in many places in Western Europe. But by looking to areas like South America and the USA we see that cultures of ethnic-religious mixing can work rather well, and we can even see many examples of how this mixing provides cultures a richness of human resources that may, for instance in the case of the USA, be one of the reasons for their strength as a nation. Different cultural inputs and different individual abilities can only enrich the total mix.

In other words, a nation of mixed ethnic and cultural inputs probably has a decisive advantage and a competitive edge in terms of human resources. This realization may *help us to fear ethnic integration less* and to learn to accept what seems to be inherent in the globalization process.

It is a question of central importance for, say, the central European countries, to relax their adherence to national ethnic purity and to ethnic and religious beliefs and to become open to the multicultural society of the future. This approach will work better in coping with the global environmental problems that the world will face in the future because to keep underdeveloped countries at a lower state of economic or technological functioning means increased conflicts and also more environmental damage that may eventually lead to wars over resources, or other serious global conflicts.

III How Spatial Systems Change

1 Today´s "Linear Centre" Around the Globe

In the previous chapter we talked about the *increased importance of time* today, as climatic changes that earlier took centuries are now taking place in just a few decades.

This stepped-up increase means that everything in the world will change at such a rate that the patterns of change can not be overlooked. *Patterns* and *megapatterns* of change will therefore be of central importance to planners and scientists as they draft a picture of *How the World will Change – with Global Warming*.

This third chapter introduces how life on Earth is organized spatially. We start, in this section, to study the *linear centre* and *ribbon of habitation* that goes around the globe.

In the following sections we will see how a new system of spatial organization is emerging. The reason for this is that global warming will make the Arctic more liveable, primarily because it is getting warmer at a much higher rate than the rest of the world or already by 4 to 7°C in this century. The change we will see first is that by 2100 the ribbon of habitation will already have widened almost to the North Pole, which will by then probably be almost ice free.

An inhabitable Arctic will therefore make the world global, as the third section of this chapter will describe.

This will effect a change from today's spatial system of a ribbon to a spatial system of a global or semi-global space in the northern hemisphere, i.e. we will, in the future, not be living on a ribbon anymore but on a space that has the form of half of a globe. This will, by gradual steps, have a profound impact on what areas will have a spatial centrality what areas will be out in the periphery etc.

In today's world centrality is enjoyed by areas closest to the linear centre that runs around the globe approximately 35 degrees north of the equator. This linear centre now goes through the Mediterranean Ocean and through the southern parts of the USA and China and the northern part of India. The habitable ribbon of the globe today extends from this linear centre about 2000 km north, reaching as far north as the latitude of Iceland, and 7000 km to the south to the latitude of the southernmost part of South America, as seen in the picture below to the left.

In earlier times the inhabitants of the areas on the rim of this ribbon of habitation suffered a great deal, both from the cold and from the distance from the central activities of the globe. However, the situation of these northern and southern regions of the world has been improving a great deal in the last 200 years.

This first great improvement came with steam-driven ships and *globalized shipping* in the 19th cen-

The widening of the ribbon of habitation towards the poles.

An inhabitable Arctic leads to a semi-global world.

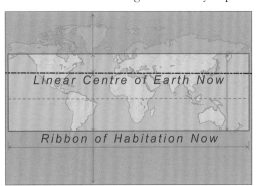

Today the linear centre runs 35° north of the equator – and today's ribbon of habitation is 10,000 km wide and 40,000 km long.

Global warming has improved the lot of the people at the cold edges of the ribbon of habitation. – Here the greening of Iceland.

These earlier isolated centres were connected by shipping and overland routes.

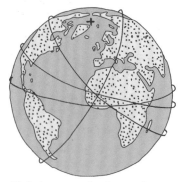

Global communication and global flights make points on the globe more equal.

Today´s language is still largely that of the flat world. Buckminster Fuller made suggestions about how to correct the terminology towards that of a global world.

tury, a process that created a somewhat connected central line around the globe where earlier large habitable areas, like the Mediterranean and China, had been relatively individual and isolated centres of habitation as travel between them had been difficult and time-consuming.

The next technological innovation that contributed to expanding the linear centre westward in North America and eastward over the Eurasian land mass was *continental railway lines,* the most famous of them being the US-Pacific Railway that opened in 1869 and the Trans-Siberian Railway that started operation in 1916.

At the end of the 19th century the development of new communication methods accounted for technological leaps in helping to connect the world. The telegraph and the telephone came first, contributing both to a general unification and also to the further development of the linear centre of the globe, with all the first ocean cables laid in an east-west direction, first over the Atlantic Ocean in 1866 and then over the Pacific in 1903.

The *wireless connection* of Marconi radio and later radio and TV broadcasts, and the extension of their reach with satellite senders, contributed to the development of global communication that was not as tied to the earlier linear infrastructures as before.

The advent of *global flights* in the 1930s at first followed rather closely the linear centre of the globe because airplanes could only cope in the most benign climatic zones over the central regions of the Earth.

Gradually, flights became possible into the sub-polar areas, and in the 1980s international flights no longer had to pass over the long distances around the whole circumference of the globe but could opt for the shortest distances over the uninhabitable polar areas. This meant that suddenly the marginal position of the northern rim of the ribbon changed dramatically: with international flights over the polar areas, the distance of any northern rim area from another northern rim area, even on the other side of the globe, had been considerably shortened.

The warming of the Arctic has meant that shipping has already increased substantially in the Arctic Ocean. With the pack ice retreating more and more from the West Siberian coast, with its access to rich supplies of resources, oil and gas, transportation from there by ship has already begun and will increase hugely in the next few decades. This area may have as large oil and gas reserves as the Persian Gulf region, which, according to current estimates, will be running out of oil in about 20 to 40 years.

By about 2050 the pack ice in the Arctic Ocean will have decreased so much that ships should be able to navigate the passage all year round. This will mean that *the huge resources* of the Arctic rim can be reached for the benefit of the whole world and transported into both the Atlantic and Pacific regions.

As the new Arctic shipping routes will be much shorter than today's global shipping routes through the Suez and Panama canals for the most important industrial nations of the world, most

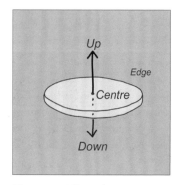

Here we realize where our understanding of up–down and centre–edge originates.

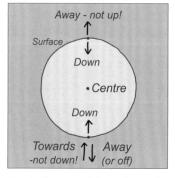

Words like up and down do not fit on a globe because up in north becomes down in the south.

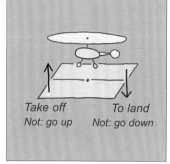

Aviation has partly adopted the new global terminology as is shown in this picture.

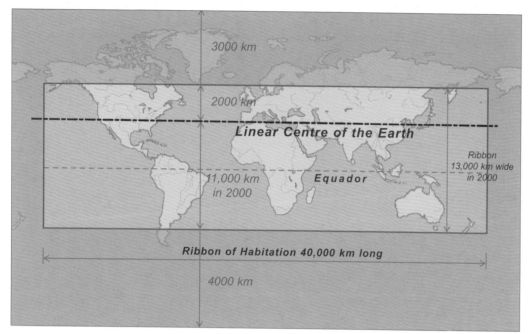

3000 km

2000 km

Linear Centre of the Earth

Ribbon 13,000 km wide in 2000

11,000 km in 2000   **Equator**

Ribbon of Habitation 40,000 km long

4000 km

The Linear Centre of Earth's habitation
now lies ca 35° north of the equator.

The Ribbon of Habitation lies on both sides of the Linear Centre. With global warming it will move further towards the poles.

With global warming the Linear Centre will move north, to the disadvantage of the South, and the Ribbon will – if the warming becomes extensive – eventually split in the middle. (See picture 5 on page 65).

This picture shows the location of today's linear centre of Earth and the length and width of of the ribbon of habitation. The picture below shows how these features wiill change in the future.

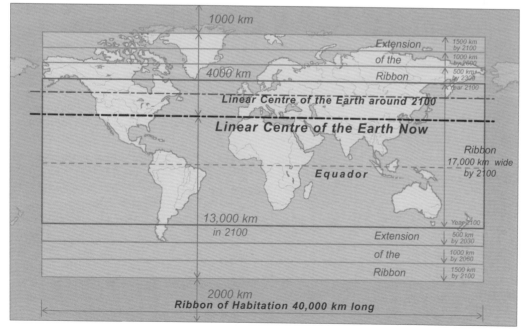

1000 km

Extension of the Ribbon

1500 km by 2100
1000 km by 2060
500 km by 2030
Year 2100

4000 km

Linear Centre of the Earth around 2100

Linear Centre of the Earth Now

Ribbon 17,000 km wide by 2100

Equator

Year 2100
500 km by 2030
1000 km by 2060
1500 km by 2100

13,000 km in 2100

Extension of the Ribbon

2000 km
Ribbon of Habitation 40,000 km long

Advances in technology and the warming of Earth's climate, have meant that the Ribbon has been expanding towards the polar areas in the last 100–200 years. The picture presents a prediction of how much the Ribbon will expand in this century, and how far north Earth's Linear Centre will move.

of the shipping between the North Atlantic and North Pacific spaces will in the future pass through the Arctic Ocean, both north of Canada and north of Siberia.

The fact that tankers and container ships are getting ever larger, in order to reduce shipping costs, means that such ships will not be able to pass through the restricted width of the Suez and Panama canals – even if they are somewhat extended – and the importance of these routes will decrease in proportion to the opening of the northern routes. With no restrictions on the size of ships, the Arctic shipping routes will therefore – in addition to the short distances – have a huge advantage over these canals – and to the routes south of Africa and S America.

Development of the Arctic shipping routes may be slowed, however, by the fact that winter travel will require icebreaker design for the ships, even though the ice will be much thinner than today. In about 100 years, however, this constraint may have disappeared almost entirely.

This whole future development of the Arctic will be dealt with in the third part of this book: *III. The Future,* especially the last chapter: *The North: The Future Area of the Globe* (page 125).

The main task of this present chapter on *How Spatial Systems Change,* is meant to clarify the difference between today's spatial system of a *linear centre* around the globe and the emerging *semi-spherical spatial system* of the northern hemisphere.

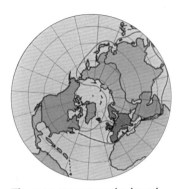

The map compares the length and location of the sea routes through the Arctic Ocean (red line) and the Suez Canal (grey line).

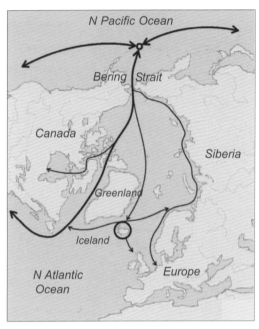

As the Arctic Ocean becomes ice free the whole year, there will be these four main shipping routes through the Arctic Ocean.

As we have already seen, the importance of the *linear centre* is declining with the advent of satellite communication, the internet and the flights over the poles that utilize these shortest distances.

The steps in the activation of the Arctic and the Arctic Rim will enhance this development. What follows from this is that the main land areas of the globe – that are actually mostly located in the northern hemisphere – will gradually gain more and more importance.

Today's *linear centre* around the globe will move north, but also in the process lose some of its importance, and instead the *geographical centre of the land mass of Earth* that is probably located at the southern end of the Ural Mountains in Russia – will gain more and more importance. The central and southern areas of the globe will, on the other hand, lose in this process, especially the areas that have already depleted their resources, like water and oil; areas, that – in addition to this – will be most hit by the growing and excessive heat and the expanding of deserts that come with global warming.

Containerships become ever bigger, making the Panama and Suez Canals ever less useful.

	Shanghai		Yokohama	
	Distance	Time	Distance	Time
	Nautical miles	Days	Nautical miles	Days
Rotterdam – Cape of Good Hope	13,889	27.6	14,506	28.8
Rotterdam – Suez Canal	9,612	19.1	11,212	22.2
Rotterdam – Northern Sea Route	8,865	17.6	7,825	15.5

Comparing travel time and shipping distances through the Suez and south of Africa, with the Northern Sea Route through the Arctic Ocean.

2 Today's "Ribbon of Habitation"

In the Stone Age when man was only a hunter and gatherer like other animals, several families often lived in shelters grouped in small communities, but there was no large-scale spatial organization of human habitation.

As humankind, however, started to develop in organized ways in terms of activities and social functions, *centres started to develop*. Sometimes these centres had religious or defensive purposes, and sometimes they were centres of trade, transportation, crafts, etc. Gradually, such tribal centres grew into clusters, often under the auspices of some secular or religious power.

A spatial feature common to most of these centres was that they were *at the gravitational centre of a geographical space* that most often approached the shape of a circle. A centre, in the beginning – about 15,000 years ago in India – was therefore actually the centre of an area, in the geometric sense of the word.

This *spatial system of the circle,* with a centre in its middle eventually led to a societal space of three spheres: The innermost sphere, the *centre,* was the space of the elite, surrounded by a second sphere of loyal supporters. The third sphere was inhabited by tribes that could be used as suppliers, and also as soldiers in times of warfare. People outside this spatial domain were deemed to be barbarians and enemies.

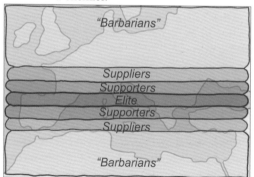

The elite lives at the linear centre of an area. Belts of social spaces develop to both sides with those outside often termed "barbarians".

This model accords well with the development of much of the Near East and of the Roman Empire, for instance, as well as Central America in the time of the Aztecs.

With the development of *transportation corridors* some of the earlier, isolated centres started to become connected into a string that can be defined as a *linear centre.* The space around this linear centre eventually evolved into becoming a *ribbon of habitation.* This linear space had the same *hierarchy of social spaces* as the circular spaces; the linear centre was the space of the elite, the belt next to the linear centre that of supporters, etc.

As the *global linear centre* around the world started to evolve with global shipping, the social implications of this new *spatial organization* of global habitation eventually became somewhat similar to the earlier social spaces.

At the beginning of this linear centre of development around the globe the lands and continents newly discovered by Western Man had the position of colonies of the countries of the old world. In due time the colonies gained independence and created their own centres of power. These "new" centres – some of them were actually old – eventually became full-fledged elements of the line or ribbon of habitation around the globe that by 1900 had already formed a circle or a belt around Earth, about 35 degrees north of the equator.

The development of this particular spatial form of the line or the ribbon therefore actually *had an equalizing impact* on the central areas of the globe. What remained, however, was that socially the spaces to the north and the south of the ribbon remained – within the worldview of those inhabiting the ribbon – spaces of little interest, inhabited by "barbarians".

Still today, this is the perception of the old world powers at Earth's linear centre. Italians, for example, do not even show Northern Europe on some of their world maps.

Global navigation demonstrated that the Earth

Earlier a centre was always a centre in an area, and the elite lived there. Around it were spheres of declining social importance.

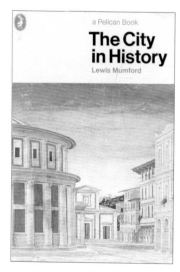

This book was the first to describe the role of urbanization in how the world developed.

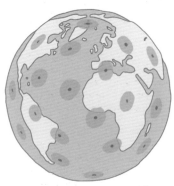

On a globe all points are of an equal importance. This is in contrast to the flat Earth view where the points in the middle of the flat area are important, but the points out on the edge much less so.

This picture shows that most of the world's landmass lies in the northern hemisphere. We, therefore, will live in a semi-global world in the future.

was not a flat area – where its centre, automatically, was of primary importance – but rather: a three-dimensional globe, where all points, *in a topological sense,* are of equal importance. (See the picture to the left).

This new spatial image of our world eventually started to change the old mental picture that still considers certain "central" areas on the globe to be of much more topological – and political and economic – importance than other areas.

The practical fact, however, remained that the top and the bottom of the globe, the polar areas, were still rather unapproachable and uninhabitable supported the perception of people in the "centre regions" of the globe that they live at the Earth's centre. Even today they can not break out of the mental prison of *their dated spatial perception of the world* and start to realize that the polar areas – a quarter of the globe's surface – provide very real potential and will eventually be of huge importance for the world, as described in this book.

Today the Arctic regions, which are already becoming of huge importance for the world, are not even shown on most world maps, and if they are shown, they appear in a very distorted and unrealistic way that makes them look very peripheral, whereas they can also in fact be viewed as

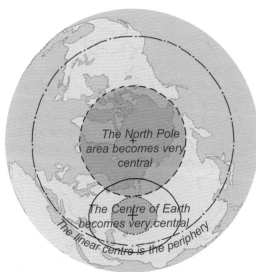

The globe viewed from the top. Here the "most distant point", the North Pole, becomes very central, but today's linear centre is out in the periphery.

equally central and important as any other areas of the globe.

Let us now continue our study of the nature of today's ribbon of habitation that embraces the globe. First of all, we need to recognize the importance of the fact that this ribbon is not merely flat with two outer ends, as it appears in most maps, but rather that it girds the globe and forms a seamless *ribbon circle.* This means, topologically, that every point at the centre line of the ribbon is spatially of equal importance. Points outside the linear centre, however, have a more reduced importance the further removed they are from the linear centre of the ribbon.

Let us now look at the globe as seen from space, with the Urals in the middle, as presented in the picture to the left. This picture shows us that most of the world's land mass lies in the northern hemisphere. And, as the globe is viewed in this way, it also dawns on us that we do not live in a *global* but rather a *semi-global world.*

It takes a lot of study to recognize what this "strange" spatial system of the semi-globe means, and will eventually mean, for the system of habitation on Earth now that the Arctic area is becoming habitable and shipping through it feasible.

The first point that is important to recognize is that the Arctic is close to the middle of the land areas of the world, which means that the Arctic Ocean can, actually, be termed the Medi-ter-ranean-ocean, the *Middle-of-the-Earth-ocean* of the globe within the spatial system of Earth in the future.

After having recognized this on a map that shows the globe from the top, we realize – to our amazement – that today's linear centre and ribbon of habitation is no longer in the middle of the future picture of the globe but rather out on its periphery!

It will, of course, take a long time until this is the picture that shows how the world is, functionally speaking. Today's global warming, however, is bringing us towards this direction of a *North Pole centred picture,* though nobody can predict as yet how extensive and over how long a period global warming will last, but the Arctic must become quite warm to let the Arctic Ocean and its envi-

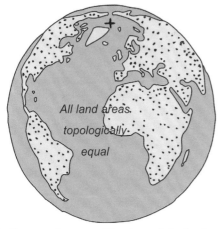

1) No central power: All points of the land areas of the globe of similar importance.

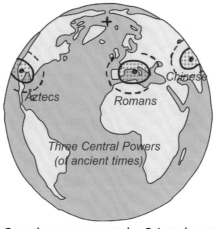

2) Central powers come to be: Points closest to a power centre are of most importance.

The development of the spatial systems of Earth in six steps.

The first step of this spatial development started with a defuse state, as there were no power structures, and humans were equal, like other organisms.

The sixth and last step: The Semi-Global System will take over as the land mass of the northern hemisphere – all the way to the Arctic Ocean – has become activated because of global warming.

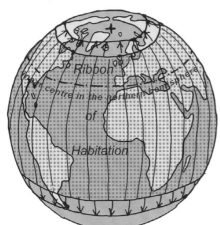

3) Ribbon of habitation: Extension to the poles. Points at the linear centre most important.

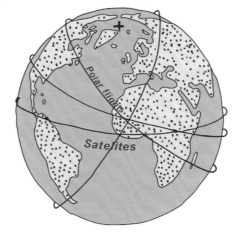

4) Polar flights and cyberspace: Points become more equal again. Peripherial areas gain most.

5) Centre of Earth becomes very warm: The ribbon splits in two. Movement towards the poles.

6) The South loses importance: The North Pole becomes a symbolic Centre of the World.

rons become the new Mediterranean Ocean and the Mediterranean area of the globe.

Let us now, for experimental purposes, draft a scenario of how the world will change with continued and excessive global warming. First of all: the Arctic will have a warm climate that opens it to use as a rich and clean environment. What follows is the development of towns, ports, roads, industries. As the whole Arctic Ocean becomes ice free, a great deal of shipping will be taking place within the Arctic itself, between the various towns and ports that will eventually surround the Arctic Ocean. In addition, some of the most important shipping routes between continents will go through the Arctic Ocean.

What will happen within the system of today's ribbon of habitation will be quite different, that is, once the Arctic has gained its rightful position as a very central place on the globe. This development has already started, and the ribbon will eventually become almost a periphery. The spaces further south of the ribbon – Africa, South America and Australia – will be still further "out of the picture", not only spatially, but also, some of them, because of excessive heat, depletion of resources and the expansion of deserts.

After this voyage of exploration into the future,

let us now return to today's situation and study how the ribbon of habitation is evolving today and how it will evolve in the nearest future.

As we have seen, the linear centre and the ribbon of habitation around the world started to develop with global shipping. Understandably, the most important towns of this epoch were the harbour cities and cities located along wide rivers navigable by ocean-going ships. The next step in the development of the global ribbon was the 19th century advent of cross-continental and intercontinental railway lines. These railway lines opened a ribbon to development on both sides – across continents. Within regions and continents other transportation systems, like canals and highways, and eventually local flights, strengthened this east-west linear development of regions and continents. The next big step, in terms of more importance of the global ribbon, was the advent in the 20th century of international flights.

All these systems of physical transportation and communication strengthened the central quality of the ribbon and its linear centre. Later, technological advances in building ships and airplanes meant that the ribbon was widened to the north and south, into areas of less benign climates.

Many northern areas are already experiencing improved climatic conditions. – Here bathing in a natural geothermal pool in Iceland.

Ships and continental trains were major agents in the development of Earth's linear centre.

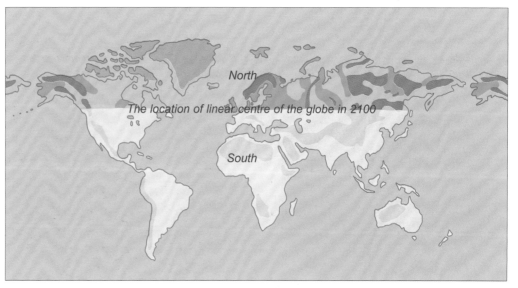

North

The location of linear centre of the globe in 2100

South

An estimation of what will be the most active areas north of the linear centre of the world in 2100. Dark green shows easily habitable low areas and corridors along rivers. Light green shows other good areas. Grey is uninhabitable areas and brown other uninhabitable areas; mountains, deserts.

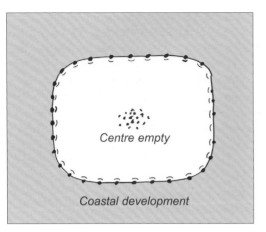

Centre empty

Coastal development

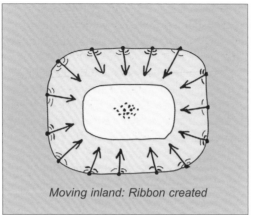

Moving inland: Ribbon created

The way a region develops, in six steps.

At first the centre is a linear centre at the coast ...

... but later the centre develops at the geographical centre of the area, as in the last picture.

1) The coast a continuous linear centre: All of its points have a similar topological strength.

2) The linear centre of the coast develops into a ribbon along the coast: Points of similar strength.

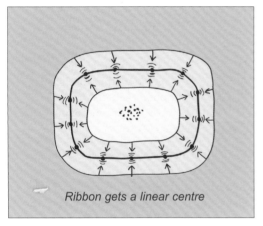

Ribbon gets a linear centre

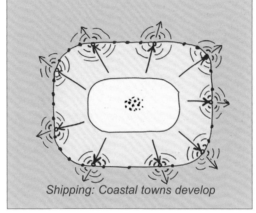

Shipping: Coastal towns develop

3) Roads develop in the middle of the ribbon: Importance of its outer edges decreases.

4) New coastal development drains interior: The coast a linear centre – with point centres.

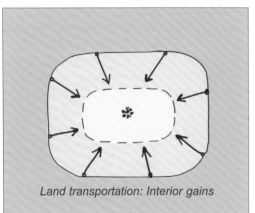

Land transportation: Interior gains

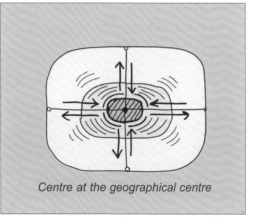

Centre at the geographical centre

5) Habitation extends towards the interior again: The coast loses some of its importance.

6) The geographical centre becomes The Centre: The coast is now of least importance.

Cyperspace started to develop in the 19th century with the advent of the telegraph and the telephone.

The phenomenon that we call *cyberspace* today started to develop in the last part of the 19th century with wireless means of communication. These new means and technologies of communication continued to develop during the 20th century with global radio, global television, and finally, a global internet. This cyberspace development has meant that areas that were at a disadvantage in terms of their distance from the linear centre of the world can now, more and more, become active members of the global community.

At the same time as *cyberspace* continued to develop, the physical means of communication and transportation on sea and in the air continued to develop further towards the poles. Both these developments thus continued to reduce the marginal position of the areas of the far North and South. The collaboration of intercontinental operations over the polar areas started with flights seeking the shortest distances by flying over the poles in the 1980s. By ca. 2050 the same will happen with global shipping in the Arctic, that is, ships will also be able to make use of the shortest distances between continents by passing over the North Pole, through the Arctic Ocean.

The elation connected to the internet and cyberspace has made scientific authors overestimate its importance, even speaking about a flat Earth, whereas in fact, it is the global form that is making all points on Earth equidistant. The authors of the metaphor of the "Flat Earth" do not recognize that physical transportation in a globalized world is essential for establishing and maintaining equal footing in the interaction between those in the various regions of the world.

Not many people of today realize how much the modern transportation systems of, say, Europe and northern America were a precondition for the collaboration of the various regions of these two continents. The same will hold true on a global scale. As shipping through the Arctic Ocean develops, it will mean a similar epoch-making development in the shipping system of the world, so much so that it will introduce a new epoch of global interaction.

But it is not only shipping that will improve global interaction with the advent of Arctic shipping. New global developments in terms of railway systems are already on the way, such as *the new Siberian railway system* from China to Narvik in Norway that is meant to be a very profitable transportation link from China to Europe. This link will also, via ships to the eastern ports of North America, be a new connection from China to the east coast of North America. The layout of this new railway line is now facilitated by the fact that the climate in Northern Russia is getting more benign so there are now better possibilities for building and operating railway lines in that region. This new railway will also connect many important industrial and resource areas within Russia, Kazakhstan and Mongolia.

These are just examples of major changes that are already under way in the world because of global warming and because of the increased globalization of trade and transportation. The new transportation infrastructures will be discussed in further detail in Chapter 4: *The Future Structure of the Globe,* which starts on page 79.

To the people in the land–locked Mongolia trains and planes are magic, because they overcome isolation.

Before ocean-going ships, areas of the high North and far South were very isolated. With them all ice-free coastal areas became reachable.

3 The Arctic will Make the World Global

The activation of the polar areas – especially that of the Arctic – will occur as the global climate continues to get warmer. In the past extreme cold has led to year-round ice cover and has primarily been the prohibiting factor for limiting the development and presence and a more extended range of biota and human activities. Global shipping, utilizing the shortest distance between continents via the Arctic Ocean, has therefore not been possible despite courageous historical attempts to find a passage.

The warming of the Arctic, on the other hand, will mean that the whole northern part of the globe – the site of most of the landmass of Earth – will be become open to a different and increasing biota and, eventually, to the development of a system of important central areas for human activities. This will lead to a spatial system of centres that, in many ways, will be different from that of the globe today.

The two basic spatial systems of the globe, the semi-spherical system of the northern hemisphere and today's middle-latitude ribbon around the globe, will co-exist for a long time to come,

but the importance of semi-spherical space will gradually be strengthened at the expense of the central ribbon space. At the same time that it loses some of its uniqueness, the ribbon space will expand to the north. As the North continues to warm it will, as a result, become spatially stronger. The importance of the South, in contrast, will weaken as, in many areas, it becomes undesirably hot for human activities.

Contributing to the decline of the spatial system of the ribbon is the fact that many of its central areas will be getting so hot that the southern half of today's ribbon will be divided from the northern half by a belt of almost uninhabitable arid areas.

In response to global warming the southern part of the then divided ribbon of habitation will also move towards the southern polar area – just as the northern part of the ribbon will be moving northward. The southern part of the ribbon, however – unlike the northern part, will lose importance since, in general, activity on the globe will be moving north. The northern hemisphere in the

As the central region of the globe gets very warm ...
... the ribbon of habitation will split, and the southern and northern hemispheres will evolve differently – as shown in the two maps below.

Four reasons why Antarctica and the southern hemisphere will not experience much development:

1) Antarctica is not a direct continuum of the land spaces of the southern hemisphere as the Southern Ocean separates it from the other continents.

2) A slow gradual movement of settlements towards the Antarctica will not be possible in the same way as in the northern hemisphere.

3) Antarctica is a polar area rather than a sub-polar area, like the Arctic Rim.

4) Antarctica and the southern hemisphere do not seem to have as many valuable resources as does the Arctic Rim, and it does not have as much land area as the Arctic.

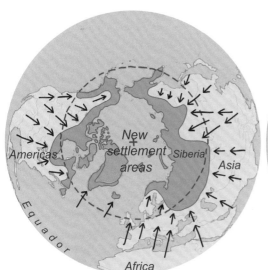

The new areas that open up with global warming in the sub-polar Arctic areas, are a direct continuum from already inhabited areas.

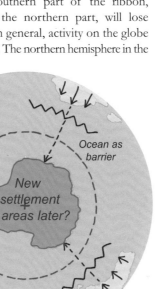

The new lands of Antarctica are separated from other continents by a wide ocean. Antarctica will also be a colder area than the lands of the Arctic.

For long it has been assumed that the Arctic would be an unchanged sanctuary, but with global warming extensive changes will happen in the area.

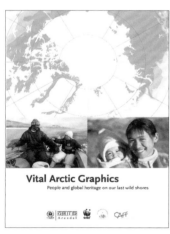

International organizations have started to gather and publish more data on the Arctic.

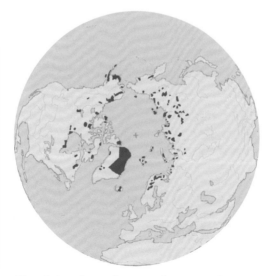

The picture shows the areas that are under some kind of environmental protection in the Arctic today.

process will become – with excessive global warming – more or less the future home of mankind.

Like the northern hemisphere, the southern one also currently has large sparsely populated areas in southern Africa, southern South America and the Antarctic. There are four reasons why the southern hemisphere will not get as much of a boost from global warming as the northern one – besides not being as large in terms of land area: 1) The land-mass of Antarctica is not a direct continuum of the land spaces of the southern hemisphere as the Southern Ocean separates it from the other continents; 2) This means that a comparable slow gradual movement of settlements towards the South Pole will not be possible in the same way that will be possible in the northern hemisphere; 3) Furthermore, Antarctica is a polar area rather than a sub-polar area, like the Arctic Rim, which means that a very long time would be required before it could become liveable and in competition with the Arctic Rim. In addition, 4) Antarctica does not seem to have as many valuable resources – such as oil and gas – as does the Arctic Rim.

These reasons, taken together, mean that it makes sense – in terms of a global planning policy – to designate large global conservation areas in the Antarctica and the Southern Ocean, rather than areas in the Arctic.

As the whole northernmost part of the globe will be getting warmer at a much higher rate than the rest of the world, it is not only the Arctic Rim and the Arctic Ocean where epoch-making occurrences will be taking place with global warming. The impact on the various sub-Arctic areas will be described in the second half of this book, but as a preview, the areas and places where most changes will occur are described later in this section.

In all this development, however, it is of primary importance that the Arctic will, with the gradual retreat of the Arctic pack ice, become open to extensive shipping.

Following are the five steps of how the pack ice will retreat and then how this will lead to five progressive steps in Arctic shipping.

The first step in the retreat of the ice from the West Siberian coast has already started during the summer months. Cargo ships are gradually going further east along the coast in the summer without the help of icebreakers. However, at present the narrow and shallow straits between the islands and the mainland make it impossible to employ large ships for transportation along this route.

The next step, a more massive retreat of the pack ice, which will mean a huge difference, will possible occur around 2030. In this period the pack ice will have retreated so far from the Siberian coast and its islands that large ships will be able to navigate through deeper waters, most probably all the way into the North Pacific Ocean.

For various reasons *the north-western Arctic shipping route* through the Canadian Archipelago will not open as fast as the north-eastern shipping route north of the Siberian coast. In the period from about 2050 to 2070 the pack ice of the Siberian half of the Arctic Ocean will have retreated almost to the North Pole in the summer. Of course it is not only summertime shipping that is important but this retreat of the ice in summer will mean that the area will only be covered by a thin coating of that year's ice in the winter. That will mean that ships comparable to today's semi-icebreaker rating will be able to pass through the thin ice in the winter without the help of ships with a full icebreaker rating.

As the ships will now be able to go directly from the Bering Strait over the North Pole into

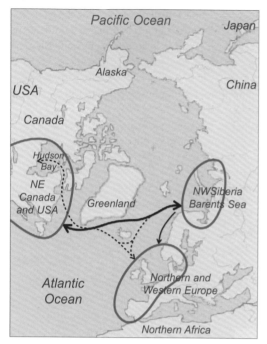

The retreat of sea ice has meant that enormous oil production is starting in the Barents Sea and NW Siberia that mostly goes to NE America.

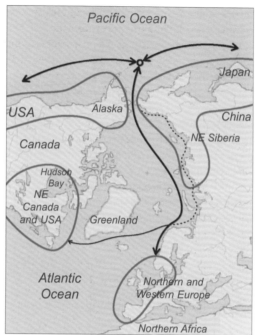

The narrow coastal Siberian shipping route will be open late summer by 2020. The route further out, for larger ships will, be open in summer by 2040.

When the ice has retreated to the North Pole this shortest route into the Atlantic will make Iceland an ideal location for depot harbours.

Shipping routes through the Canadian Archipelago will be the shortest routes between the Pacific and the east coast of Canada and the USA.

The sign of the two year chairmanship of Russia of the Arctic Council.

the Atlantic, this sea route, first of all, will be highly competitive as it is much shorter than the curved sea route along the Siberian Coast. This shorter sea route will primarily mean an advantage for the ships that will be coming from the Pacific and headed for the eastern coast of North America. These ships will enter the Atlantic through the Denmark Strait between Greenland and Iceland, thus increasing the importance of the ports in north-eastern Canada.

Hudson Bay will have also become largely free of ice and the importance of Churchill will also increase. The great advantage of ships being able to navigate freely in Hudson Bay – which is about half the size of the Mediterranean Sea – is that this body of water, which reaches into the central area of Canada, will facilitate transportation to and from central Canada. Previously Churchill has been largely developed as a railhead and port for some exportation of wheat and other grains from the Canadian plains, but the ice has been a big problem. More recently Churchill has also become a tourist centre. With global warming Churchill, like other northern areas, will continue to develop.

Once the shipping lanes through the Canadian Archipelago are open most of the year in late 21st century the development of N Canada will take off.

Many scientists predict that the next steps in the development of the Arctic sea routes will be the opening of routes through the archipelago in Northern Canada.

The fifth and last step in the opening up of shipping activity in the Arctic Ocean will probably happen around 2100. By then most of the Arctic Ocean will probably be ice free, even in winter. By this time, most of the Arctic Rim will have developed substantially in terms of industry, mining and oil and gas exploitation.

An ice-free Arctic Ocean that approaches a circular space will allow a multitude of trans-Arctic Ocean shipping lanes between the various towns and ports of the Arctic Rim. By 2050 transportation from the rim of the Arctic area into the depths of Northern Canada, Alaska and Siberia will already have developed, continuing to open more new areas filled with tremendous resources. This will, at first, happen with shipping along the rivers that, at first, will only be ice-free during the summer months.

This prediction of how new settlement develops from the ocean coast, and then gradually

Hudson Bay is about half the size of the Mediterr-anean Sea. Once it has become ice free the development of N Canada will take off.

inland via rivers, is the way new areas have been settled earlier in most regions of the globe. At the same time as the inland areas become developed, shipping from the Arctic Rim into the two largest world oceans, the Pacific and the Atlantic, will also have increased greatly.

Of course, it is not only the Arctic area itself that will benefit from the warmer climate and the increase in global shipping in the area. The area that is probably going to be most impacted, and that very soon, is the area on both sides of the narrow *Bering Strait* that connects the Arctic Ocean with the Pacific. To the west of this strait is the Russian Kamchatka Peninsula and to the east the US state of Alaska. This strait will eventually also become of *huge geopolitical importance,* comparable to the Strait of Gibraltar in earlier periods of world history.

Kamchatka Peninsula, which is about the size of Great Britain, is very rich in resources. To the west of it lies the Sea of Okhotsk, which is about the size of the Mediterranean. Harbours in this area have already started to evolve, but the harbour that is most likely to thrive is Petropavlosk.

A second very interesting point is that China has nowhere access to the Pacific from its northern regions, only from its central coastline, which

with northern trends, is very negative to China.

Alaska is located on the eastern side of the Bering Strait. The added shipping that will navigate the Bering Strait will increase the importance of Alaska's western and southern sea coasts, and most probably also the northern coast, which is the site of huge oil and gas reserves.

South of the Bering Strait are the Aleutians, a curved chain of islands that is the extension of the Alaskan Peninsula, where several harbours are located. Ships comming out of the Bering Strait and sailing west to Canada and the USA, or east to China and Japan, will be passing by the peninsula and through the archipelago giving some of the harbours the potential to become important supply depots for Arctic shipping.

The ocean area where the North Atlantic opens into the Arctic Ocean is much wider than the Bering Strait and consists of three main "gates". The westernmost gate is the *Davis Strait* between Canada and Greenland, about 340 km wide. The *Denmark Strait* between Greenland and Iceland (286 km wide) will, in due time become important

On the symbol of the UN the Earth is shown with the Arctic in the middle. It actually shows, the worldview of the future – that of one unified world.

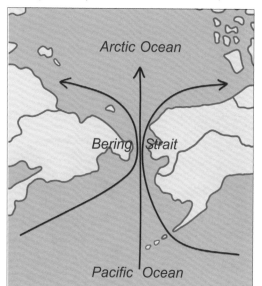

All ships that will go from the Pacific, over the Arctic Ocean, into the Atlantic space will have to pass through the narrow Bering Strait.

Mao agreed that the USSR would close China from the Pacific in the north, so the USA would have to enter the USSR for invading northern China.

A floating World Capital on the North Pole would have strong symbolism by being located in three continents – and by belonging to all time zones.

A New World

The Arctic from above
Is like a new planet.
Frost and isolation
Made it unknown.

A warm Arctic
Is a new paradise;
A place to escape to,
From excessive heat.

Exploration and settlement,
Once a pride -
Have soured
From bad conduct.

Arctic settlement
Has to be planned
To make sure
It is sustained.

for Arctic shipping. The third and the widest gate – and the most ice free – is *the gate between Iceland and Norway,* which is about 800 km wide. Here the sea lane divides into two: a lane east of the UK to the northern border of Europe and another one passing west of the UK into the Atlantic. This gate between Iceland and Norway will, for a long time, be the most important gate into the North Atlantic area.

The importance of these gates has already started to become apparent with the beginning of enormous *oil and gas transportation* from Western Siberia to north-east America and Northern Europe.

During World War II, and later during the Cold War that followed, the gate between Greenland, Iceland and the UK, called the *GIUK gate,* was of strategic importance. During the war all the ships coming from the USA with supplies to arm and maintain Russia as an eastern front in the battle against the Germans came through this gate. *Iceland,* which was occupied by the Allies, played a huge role because of its location in the middle of this gate.

From Iceland the Atlantic convoys received support and protection from battleships and airbases. After the Second World War the former Allies, the USA and the Soviet Union, became bitterest enemies. The only route for the Soviet-Russian navy to enter the Atlantic was the GIUK gate that, therefore, became a very important first line of defence for the Americans. The Americans therefore signed a bilateral defence treaty with Iceland and were allowed to increase

the size of the war-time military base in Keflavík. The Americans also set up an extensive system of submarine surveillance and of radar and air surveillance with stations in Greenland, Iceland and the UK. This system meant that the Soviet-Russian nuclear submarines could not approach the USA close enough to launch missiles without being detected.

Because of the huge oil and gas transports which will go through this gate in the very near future, this gate will *regain huge geopolitical importance.* It is therefore rather surprising that the US government practically cancelled the bilateral agreement guaranteeing the defence of Iceland in the spring of 2006.

This act of the US government is a clear demonstration of the *lack of long-term thinking* and *a lack of understanding* of how important the Arctic, Arctic oil reserves and other resources, and Arctic shipping routes will be in the future. This act was a deep insult to the Icelandic nation and the USA will have a hard time regaining its earlier position in this island nation with its strategic emplacement in the middle of the GIUK gate.

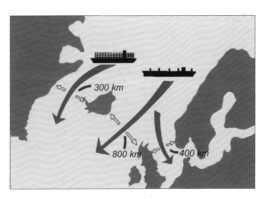

The GIUK gates are the entrance to the Arctic space, and the route for ship traffic between the Pacific and Atlantic spaces.

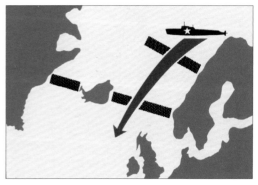

During the Cold War Iceland was the main guardian of the USA – in the middle of the the GIUK gate – against an invasion of the Sovet fleet.

4 Not a "Ribbon World" but a "Global World"

In order to carry out the task of this book to describe the various emerging changes in the world – mostly propelled by global warming – and what they will mean for the evolution of human settlement in the future, this book is divided into three parts. The first part, *Global Warming,* describes in general what present and future global warming means in terms of changes in natural and human environments. The second part is identified with a single word: *Patterns.*

This second part is divided into two chapters: *How Patterns Change* and *How Spatial Systems Change.* This is the theoretical part of the book presenting the main dynamic patterns at work in changing the world.

The term *pattern* is related to the word *trend,* which signifies a trajectory or indication of how something is happening. Sometimes the study of trends reaches far back in history, identifying *historic trends* for example in terms of technology and social development.

The most famous student of historic trends was *Darwin,* who, through his study of organisms, came to realize that the forms he was observing had evolved from earlier forms. His theory of evolution is therefore a theory of *trends or patterns,* which explicate the ways in which organisms evolve.

Let us now retrace – using the Darwinian model – how we, in this book, have been studying patterns

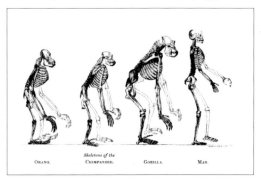

Darwin's study is the most famous study into evolutionary patterns. Settlement patterns also evolve in steps, as described in this book.

in a somewhat similar way. This book, however, studies patterns on a much broader scale, i.e. how life and habitation on Earth have evolved in general, leading to nothing less than a *theory of spatial global evolution.*

It is important to realize that the work in this book is aimed not only at understanding the principles and patterns of how things have evolved in the past, as with Darwin, but that the patterns should also be used to learn how things are *likely to evolve in the future.*

Useful methods to study trends have themselves evolved a great deal in past decades. To visualize what trend studies may incorporate, it is useful to review what trend studies within the fashion industry involve. The first step is the study of new and emerging developments in materials and methods. Fashion trend studies also review the general *historical patterns* in how fashion evolves – the length of skirts, the type of neckwear, changes in men's and women's hats – and predictions and recommendations are based on the observations.

Within planning and technology many methods have been developed to create a picture of where future developments will go. The *Delphi method* is, for instance, used in many fields. It involves expert surveying of a given field followed by a study of the data to find patterns and indicators of where things are likely to be headed.

As already has become clear, this book tries to study trends and patterns in terms of how certain aspects in our lives on Earth have evolved and changed. The first section of *Part II: Patterns* identifies indicators or *parameters* as to how our worldview is changing. The most prominent indicator is the *increased importance of active time.* This means that we are moving from a *static world* towards *a world of time and change.*

The second section of chapter 1: *How Global Patterns Change,* is called *Changes in Weather Patterns are Nothing New.* The current warming of the globe means that various weather patterns will change in the next few decades, but past changes can inform us about the possibilities of what this will mean.

Changes in weather patterns lead to *changes in the patterns of the natural world* as well as *changes in the patterns of human habitation* – on a local, as well as on a global scale.

It is also useful to recapulate that the first sections of the present chapter on *spatial systems* dealt with patterns that are of relevance for how the world will change and also examined the evolution and shifts in spatial systems as they relate to human habitation on Earth. One of the sections started by studying how centres of habitation started to evolve as point centres. With the development of transportation lines, linear centres started to evolve, first on a local scale, but eventually on a global scale – with global shipping – as a *linear centre around the globe.*

A space on both sides of this line gradually evolved to become an interconnected *ribbon of global habitation* that goes around Earth. In terms of topology, this is a two dimensional space connected to form a circle or a cylinder, a spatial system that this book calls: a "ribbon world".

The form and the topology of the space in which people live greatly influence physical functions and activities as well as social systems and people's perceptions, not only of the space in which they operate but also of themselves. As we start to study the centre line of the ribbon of habitation, we discover that its central location has many advantages in terms of physical functions, for instance in terms

of transportation, because it is located at the heart of the settlement area. The outer borders of the ribbon, *the periphery,* on the other hand, are obviously spatially disadvantaged.

The form of the ribbon also translates into something similar in terms of social functions and human perception; the areas at the centre of the ribbon are of greatest social importance, in terms of the global community, with people out on the periphery perceived by the people in the middle as inferior and uninteresting.

Of course, the people who live in the high north and south of the globe are not as advanced and refined as the old cultures which have occupied the areas of benign climate at the linear centre of the globe, though this does not really mean that they don't deserve as much respect.

The first two sections of this present chapter described the nature and the evolution of the "linear centre" and the "ribbon" of the global habitat, which together form today's spatial system of a "ribbon world".

In contrast, the last section, the third section, of this chapter on global, spatial systems has described how *The Arctic will make the World Global.*

Before we start to review the characteristics of the spatial system of a *global world* – that has already started to evolve with the activation of the Arctic – it is important to examine the characteristics of these two contrastive systems, the *ribbon world* and the *global world,* in such a way that their nature and their consequences, in terms of how areas and people

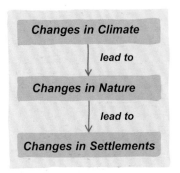

Logical sequence in the changes that come with global warming.

Three spatial systems in the history of the world.

How cultures understand the world is very much based on the spatial dispositions that are present in the ruling cosmology of their times.

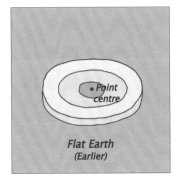

Flat Earth
(Earlier)

Within this worldview countries at the Centre were more important than those at the edge.

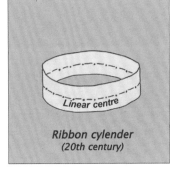

Ribbon cylender
(20th century)

The centre has become a linear centre with all points equal. But there is still a periphery.

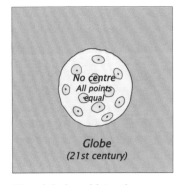

Globe
(21st century)

The global worldview has an equalizing effect for nations because all points are equal.

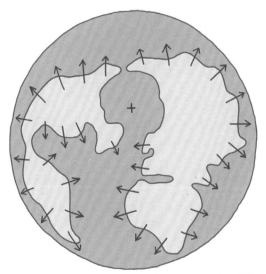

Today's world is composed of two main "islands": The Americas and Eurasia. They have no common focus that can provide a sense of oneness.

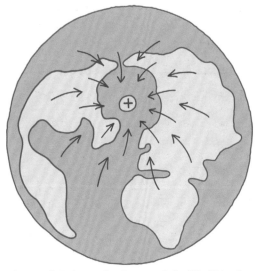

The North Pole as the Centre of the World in the future – and a floating World Capital there – can give focus and increase solidarity on Earth.

Today's worldview is a mixture of the globe and ribbon world-views.

The semi-sphere will be the worldview of the 22nd century, if the warming continues.

operate here on Earth, can be compared and understood in the best possible way.

In addition to the present studies of geometric characteristics of worldviews, the shift from today's world of two "islands", the Americas and Eurasia, and the future world that is characterized by a circular form of a landmass that surrounds the future "Middle-of-the-World-Ocean", the Arctic Ocean, should be mentioned. The importance of this shift becomes apparent in the two pictures above, These two pictures the author created in co-operation with Manuela Campanini.

A profound understanding of the two spatial global systems is absolutely essential for being able to understand what consequences the gradual shift from a *ribbon world* to a *global world* is going to mean in terms of how the world society is going to function within the emerging new spatial system of a *global world*.

Let's now start this description of the spatial system of the emerging *global world* with a correction: The world of the future – practically speaking – will not be global but rather *semi-global,* that is, a *world of the northern hemisphere.*

There are several reasons why this will be so. *First* and most important is the fact that most of the land mass of the Earth is located in the northern hemisphere. *Secondly,* the linear centre of the globe, and

therefore the most important areas of the globe for humans, are located in the northern hemisphere. *Third,* the Arctic will warm much more than Antarctica, and the Arctic will – for several reasons – be of far more importance than Antarctica in the future development of human activities. *Fourth,* excessive heat in the central regions of the world will mean that this belt around Earth will be of diminished global importance and will eventually function as a separation between the northern and southern hemispheres, which will further enhance the importance of the North and be a considerable disadvantage for the South.

The four main reasons why the world of the future will not be global but rather semi-global, that is, a world of the northern hemisphere:

1) Most of the landmass of the Earth is located in the northern hemisphere.

2) The linear centre of the globe, and the most important areas of the globe, are located in the northern hemisphere.

3) The Arctic will warm much more than Antarctica, and the Arctic will be of far more importance.

4) Excessive heat in the central regions of the world will eventually function as a separation between the northern and southern hemispheres.

Characteristics of the shift from a ribbon world to a semi-spherical world.

The picture to the left:
Within the ribbon spatial system, the North Pole is the point furthest away from the linear centre of habitation as it is today.

The picture to the right:
Within this future semi-sphere of habitation, the North Pole is the centre – and today's central areas have become the edge. This semi-sphere is going to be the spatial system of the 22nd century.

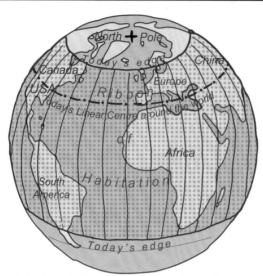

Today's worldview is that of the ribbon, but the influence of the spatial characteristics of the sphere are becoming ever stronger.

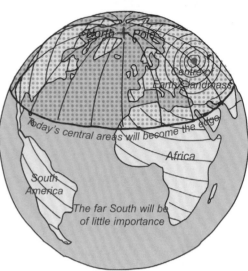

The transfer from the ribbon world to the global world will not be completed because the semi-global world shown above, will take over.

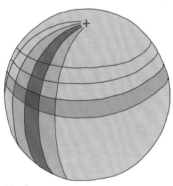

Today
relationships among cultures are primarily along horizontal belts: Northerners, Southerners, etc.

In the future
relationships along vertical zones will gain importance with the North Pole as a centre of the world – and also because vertical travel within the same time zone is easier on people than across the time zones.

These four reasons – together with an activation of the Arctic with shipping and resource exploitation – will mean that the world of the future will – practically speaking – be a *semi-global world of the northern hemisphere*.

This world of the future, with its semi-spherical shape as a stage, will function in ways that are very different from the ways activities and interactions function on the cylindrical ribbon of today.

The contrast between the two figures on this page – which represent the two spatial systems – says more than many words. On the picture to the left, the centre of the globe runs in a linear belt around the globe, situated in the middle of the ribbon of habitation. The two polar areas are of absolutely no importance, and the North Pole, even though it is not far from being the centre of the whole land mass of Earth, is the point on the globe that is politically and socially as well as geographically farthest from the linear centre of Earth as it is, and functions, today.

The picture on the right shows – in contrast – the semi-global world that will come to exist with the activation of the Arctic. Here the Arctic region is the centre of the world, with today's linear centre relegated to the periphery!

How long the shift from the *ribbon world* of today to the semi-global world of the future will

take depends primarily on how fast the Arctic warms. If the warming stops by, say 2100, it will not have warmed enough for a total shift to an Arctic-centred global world to have taken place.

Until this shift happens, the world population will live coexisting in two spatial systems that in many ways will be competing for influence. This state of coexistence, and undoubtedly a resulting competition, will of course be the state that will begin to emerge as the Arctic starts to become warmer and more activated in the next few decades.

If global warming continues beyond 2100, and if the warming becomes excessive, the Arctic will bask in a very benign climate in the 22nd century. With this same rate of warming the whole central region of the world could become almost uninhabitable because of excessive heat, desertification and lack of water.

In this case the world population will understand how lucky it is that the Arctic region exists because it will be the area where people can escape to from the excessive and devastating heat of the world's central regions.

4 The Future Structure of the Globe

1 Transportation Structures

In the first half of this book we studied, in a rather theoretical way, how the climate of Earth is changing. We also studied how the various patterns and spatial systems of the globe change with changed conditions. In the second half of the book that starts here, we try to draft a more detailed picture of the future. This part is divided into three chapters, and the present one, *The Future Structure of the Globe* outlines some of the most important physical aspects of how the globe will look in the future. Once this picture has been drafted we can proceed to next chapter: *Impacts on a Global Scale*, where we access the picture drafted of the future in this chapter. Here we will summarize what areas of the globe will gain or lose with these changes and we will access the impact on the global economy as well as on geopolitics.

Once we have this rather clear idea of how the world of the future would look and what the impact on various regions, the economy and politics will be, we proceed to the 6th and last chapter of the book: *The North: The Future Area of the Globe*. The reader will by now have understood rather well that it is the Arctic that will experience the most global warming were most changes will take place. This goes for instance for transportation, accessibility of resources and changed life styles.

There are of course many areas in the central regions of the globe that will be very negatively influenced with global warming because of drought and depletion of various resources, most importantly of water. It is primarily in the North and in the Arctic and sub-Arctic regions of the globe where these problem areas will be *substituted for by new areas* that will appear in a pristine form from below the present snow and ice.

The last chapter of the book draws a picture of the warm Arctic of the future with short shipping lanes between the Atlantic and Pacific oceans, which will make the whole Arctic rim on both the American and Eurasian sides readily accessible – an area full of resources. Enormous oil and gas reserves have been discovered in these areas so that the pressure is already on to start intensive development in some of the Arctic areas, for instance drilling for oil in the Barents Sea and western Siberia and in northern Canada and Alaska.

The author of this book is a planner and there is a certain methodology that the planner has at his disposal to draft a picture of the future based on various types of data. Some of them are provided by maps, others by studies of climatic conditions. It does not really matter if the area that is to be studied by a planner is small or large like in this case the whole globe!

Today there already exist *planning offices for continents* and for large entities like the European Union, the United States and the Russian Federation. The scheme that is usually first developed in such planning bureaus is *an overall system of infrastructures*. This is mainly because infrastructures usually start to develop regionally without an overall plan. There therefore has to come a time in the development of a region where a planning institution, high in the hierarchy of government, has to be established to co-ordinate and simplify these pre-existing infrastructures and to plan and develop additional ones.

A point in case is how *the infrastructure of high speed railways in Europe* has been centrally planned in Brussels, which has led to including, rethinking and redesigning all the other transportation structures in Europe. The result is a total transportation infrastructure that will fit together as a co-ordinated and integrated transportation plan for the whole of Western Europe.

On a global level not much planning has been done as yet in terms of infrastructure on an institutional level. However, we probably very soon will arrive at the stage of *globalization* where it would be a wise move to create such an office, an office that has the mission of creating or drafting a picture of the most sensible future transportation structures for the whole globe.

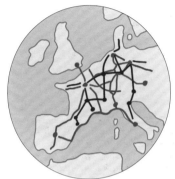

The transportation systems of Europe are being integrated so they will intercorrect.

In order to help the reader to understand the argumentation and the work process of the upcoming chapters, it is necessary to give an introduction to *the main steps* as to *how a plan is made*. The work of the planner divides into two parts: *The first part is preparatory work* to be able to draft a picture of various features of the future and is divided into four steps. *The first step* is the mapping of today's structures. *The second step* is to make maps that show good areas and bad areas, meaning: to show the areas that are best fit for settlement and areas that should be avoided. *The third step* is to study how some main basic conditions will change, are changing or will change in the future.

These include, typically, various climatic factors and socio-economic developments. *The fourth step* is to study how spatial structures and systems have evolved. This has been the main subject of some of the chapters in the earlier part of the book and the page numbers will be referred to in this section.

Once these foundations have been established it becomes much easier to draft a picture of the *planning elements* of the future, in our case of the globe as a whole. *Here we again have the four steps* of the work process, but now we start to refer to future developments and conditions. *Step one* is to make maps of the global infrastructures in the future, as shown in this section. *Step two* is to make maps that show which are likely to be the good and bad areas of the globe in the future; this step will mostly be dealt with in the next section, starting on page 85. *The third step* is to outline the most basic changes in conditions in the future, mainly climatic changes. This has already been described earlier, for instance in the sections beginning on pages 13 and 19. Finally *the fourth and last step* is to describe the spatial patterns and systems of the future – in this latter part of the book – in a more detailed way than was done in the earlier half, primarily in the sections that start on pages 63 and 75.

Let us now start to study the most important *characteristics of our physical world* in terms of *transportation structures*. On this spread we have maps of the *main barriers* in the development of the two most important infrastructures of the globe: the *global shipping* and *railway systems*. These two systems will be altered in the future by the changed physical conditions resulting from global warming. The system of global flights and telecommu-

Mountain ridges are huge barriers to land travel. Mountain passes offer some possibilities.

The Panama and Suez Canals opened ways through continents. But they are narrow.

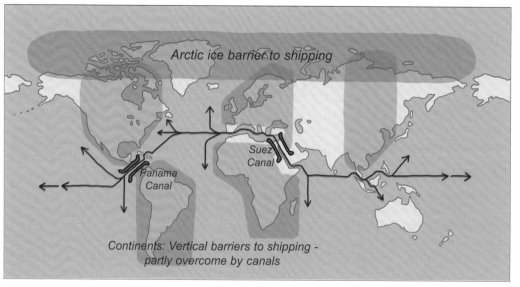

Arctic ice barrier to shipping

Panama Canal

Suez Canal

Continents: Vertical barriers to shipping – partly overcome by canals

At the beginning of global shipping the main barriers were the continents that divided the world oceans. This barrier was partly overcome by the Suez and Panama canals and by the discovery of the routes south of Africa and S America. The routes north of the continents were closed by sea ice.

nications are not studied here because their basic premises will not change that much.

The picture on the left page shows today's main barriers to constraint-free world shipping systems, and the picture on the right page shows the main barriers to the world railway systems. On both the maps, main obstacles that have to be overcome in these two modes of transportation are shown in pale brown or blue. In both cases, it is *the high North* with its permanent layer of snow and ice that is *the main obstacle today*. The present ice cover is prohibitive for both the transportation modes and therefore leads to the absence of shipping or railway connections along the whole high North and the Arctic area. In terms of other areas of the globe, it is mainly the vertical form of the continents that has been a major hindrance between the world oceans and continents. About a hundred years ago, a huge step was reached in transcending these barriers: the building of the Suez and Panama Canals.

As we see in the map on the right page, the *main barriers* to continental railway connections have been *the huge vertical mountain ranges* of North and South America, and in Asia the horizontal mountain range reaching from the Caspian Sea to China. In Africa it is mostly the Sahara desert that is a barrier to building railways.

Let us now proceed to the next two sets of pictures *in the next spread*, showing *ideas on global shipping and railway systems for the future* as snow and ice has mostly retreated from the Arctic and sub-Arctic areas. With this the main preconditions for the two systems will have changed dramatically. The retreat of snow and ice in the Arctic areas, will then open up shipping lanes across the Arctic Ocean and a better possibility for vertical north-south railway lines in the northern areas of the globe.

Another novelty regarding global transportation – that may well become a reality in the future – is the introduction of *high-speed trains for connecting distant points of the globe*. We are here dealing with a system of submarine or underground railway tunnels where the trains will be "floating" on magnetic cushions (see the lower picture in the column). Engineers have already developed this possible future transportation system *for connecting continents*. The main benefit of these trains is based on the lack of air in the tubes and the magnetic cushions, both of which minimize friction and the effort of travel. This means that very high speeds can be

Trains with snow ploughs offer fine transport facilities in flat northern terrains.

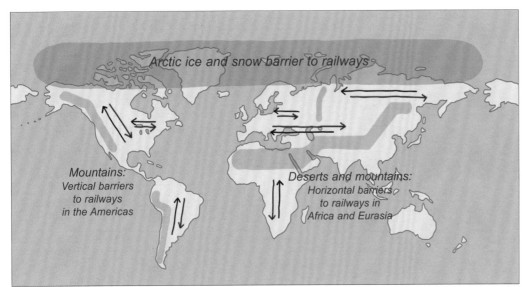

The main barriers to continental railways in the Americas are the north-south mountain ranges. In Africa and EurAsia these barriers, that are mostly the Sahara and the Himalayas, have a west-east direction making rail transportation between south and north difficult in many regions.

A system of vacuum tubes for high-speed trains would allow for connections as fast as jets.

reached and the energy costs will be very low. Such a train system has been drafted to cross the Atlantic.

The author of this book suggests that a global network of such high-speed trains can be created in the future, partly replacing the very energy consuming and polluting system of air travel.

Another great benefit of the system will be that the *trains can arrive in the heart of a settlement area,* in railway stations within the metropolitan areas, as with trains today, or at a hub of integrated international transportation such as major international airports. In this way transfer to another transportation medium takes very little time, is very convenient and also lessens travel time. As can be seen in the picture on the right page, the author suggests *a triangular global system* connecting the three main continents. The author suggests that the central train stations should be *Beijing* in south-east Asia, *Berlin* in Europe, and *Chicago* in North America. The reason for the selection of these cities is that, with the exception of Beijing, they are at the heart of geographic centres of continents, but all of them are placed in the heart of important metropolitan areas.

People might ask why Chicago and not New York. One of the answers is that New York is out on the edge at the eastern sea board of the American Continent whereas Chicago is *closer to the geographical centre of the continent.* Because of future migration to the north with global warming Chicago also has an advantage. In the future Chicago will not be so much a northern city as it

The creation of circles in transportation systems *has a great equalizing effect; every area has the possibillity of trading in both directions. And if somthing happenes on the one route – like in times of war – the other route is open.*

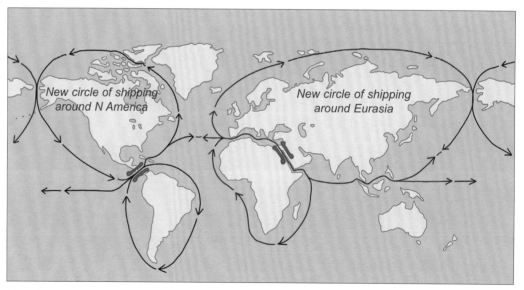

A view from the North gives a much better idea of the nature of the future shipping system – especially in the Arctic region.

Many container ships have become so large that they can not pass canals.

The shipping system of the future is four circles and connections across the Atlantic and the Pacific. In the lay-out of transportation systems it is very important that they form circles. In this way routes are not cul-de-sacs, with places at the ends highly disadvantaged, as in the North today.

is today, but rather a central city in the future settlement areas of North America. As today, shipping through the Great Lakes and train connections will continue to enhance Chicago's potential as a trading hub. Similarly, Berlin is also today a rather northern city; Munich, for instance, would be closer to the centre of gravity of Europe as it is today. But with an increased population in northern Europe, Scandinavia and Russia, *Berlin* will gradually develop into becoming more of a *geographical centre* for the European continent.

The high-speed trains would primarily be fit for transport of passengers and light cargo, but railways and ships will continue to be the most economical way of transporting heavy goods that need not go very fast between destinations. *Increased globalization of trade* will mean that shipping of goods between areas will increase still further. Actually, it is the inexpensive shipping costs that have made it possible to utilize the inexpensive labour available in the developing countries to a high degree for producing goods for the entire globe, to the benefit of both the developing countries and the consumers in the more developed countries. And, of course, these transportation systems will, as they have in the past, have a great equalizing effect among nations.

The world is fast becoming one big market place and one big labour market, where it will no longer be possible to remain within fortresses constructed and protected by trade barriers or technical barriers, as is the tendency of the USA and the European Union today. If these unions continue their isolationist policy in order to protect their internal markets, their industries will become less competitive, which will gradually lower their living standards because of higher prices for, say, agricultural products, as in Europe today. One idea behind these unions is to protect their own work places, but in due time this will mean that the protection measures will result in placing themselves at a lower stage of the industrial ladder. The only way ahead, as always, *is to accept change,* go with the changes, and keep abreast and ahead of development, as possible. Those that are the fastest in adjusting to the new and changed conditions will be best off.

One of the central theses of this book is that

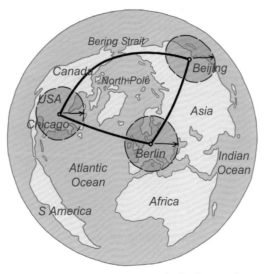

Three possible main centres of a high-speed global railway system, running through vacuum tubes. This system could partly replace air travel.

today's horizontal structure of transportation and habitation will decrease in the future. Now, with this study of future transportation systems in this section, we see that they are becoming free of the constraints of northern cold and ice so they will mor easily be able, in the future, to run vertically or north-south, across the horizontal belts. This will be one of the factors contributing to the

The areas that will benefit most from a closeness to the three main centres of a global train.

Trans-continental railways (green) connect the east and west coasts of N America and Eurasia. Two more northern connections will come soon.

The Kremlin was earlier a symbol of a very closed society. Now its co-operation with the West is of utmost importance

development of the northern areas of the globe and will contribute to the lessened importance of today's horizontal, linear centre around the globe: We are moving away from the spatial structure of the ribbon towards the direction of an *economic and habitational semi-sphere* of the northern hemisphere.

Of course, other global structures will importantly affect how the world develops. Obviously, political structures are key. Today we have mostly *four blocs: the Western countries* of Western Europe, the USA and Australia; secondly *the Russian Federation;* then *south-east Asia and the developing world;* and fourthly *the Muslim world.* Today there are huge hostilities, primarily among the Muslim and Western worlds, that will prevent these two cultures from merging and co-operating, meaning that the Muslim world will rather seek to influence south-east Asia and Africa, mainly through its oil.

There are signs that the former enemy of the West, Russia, is going to be a very close collaborator of Western countries in the future. The political differences between communism and capitalism have no longer kept these two blocs apart since the fall of the USSR in 1989 and 1990. As often before in history it is the common enemy, the Muslim world, and common commercial interests that tie the two former antagonists together.

The big problems that the West and the USA have in the *Persian Gulf region* have led to the fact that the West is trying to assist the Russians in the development of their huge oil and gas reserves. Global warming and the retreat of sea ice in the *Barents Sea* and along the coast of *western Siberia* is the key for creating a Western oil partnership with northern Norway and Siberia.

North European countries like the Netherlands and Britain also have a great deal of interest in this because the oil resources of the *North Sea* are approaching their end in the foreseeable future. Because of a better climate, increased searching for oil is also taking place in other northern areas, for instance off the Faroe Islands, at Jan Mayen Island, north-east of Iceland, and along the west coast of Greenland. The Canadian archipelago offers other potential-

ly great oil fields that will be more accessible with global warming and the retreat of the pack ice; however, because of a slower retreat of the ice there, the Canadian area will be accessible somewhat later than western Siberia.

Today a large part of the oil and gas production of these Siberian fields goes overland to Europe, and Russian gas is today supplying Western Europe with a large percentage of its energy needs. Recently also a new pipeline has been built to China.

Liquefaction of gas is costly, but it is on the increase and will be part of the new northern development. The big benefit of the gas is that it provides much cleaner energy than does oil. However, in spite of these newly discovered oil and gas fields, coal, nuclear energy, and renewable energy resources will also be developed. The problem with coal, although it is inexpensive, is how dirty it is.

The problem with nuclear energy is that politically its development is connected with the possibility of creating nuclear bombs. The big problems with the so-called renewables is that they have severe environmental impacts, contrary to what many people think, for example in terms of noise and visual impacts. Also, most of them are very expensive, so this kind of energy will probably not really take off until, and if, oil prices go still further up permanently.

The new oil production that is now being developed in the North has mostly come about for *geopolitical reasons.* The problems that the West has with the Muslim countries of the Persian Gulf, as well as some other oil producing developing countries, could mean that so much investment is being made in the northern areas for political reasons that there will be *an overproduction of oil early in this century.* This would mean that development of alternative energy resources would not develop further substantially for the time being, leading to increased production of greenhouse gasses and therefore to still more global warming.

To be able to drill for oil out in the sea opened up huge areas for oil and gas extraction.

2 Good Areas – Bad Areas

One of the main tools of the planning profession is the *evaluation of areas in terms of negative and positive aspects*. As this and other studies have been carried out, a plan is created based on system ideas as well as on the qualities that have been evaluated to be inherent in the area. Understandably, one of the basic goals of a plan is *to avoid the negative aspects* of an area and *make efficient use of the good qualities*. Most often this evaluation work is presented on several separate maps. Earlier such maps were drawn on transparencies in order to be able to view them together, but now such work is carried out on a computer. There these maps can be viewed separately or laid over each other so the sum of the good aspects or the bad aspects in question are revealed on a single map.

The planner usually starts by mapping the negative aspects of an area, the so-called *constraints to development*. This is because the constraints maps show areas that are negative or of little interest. To start with the constraints makes it easier to see which spaces one needs to focus on in the assessment of the positive qualities inherent in the area. The first step in the process of creating evaluation maps is *to make a list of the positive and negative features* relevant for the area and the plan. Such lists vary a great deal in terms of what scale the planning takes place.

If we are working with a small area, the depth of soil can be a very important feature because to build low buildings on deep soil is very costly. Coming up to a regional scale of planning, features like closeness to airports and harbours become positive features, whereas areas close to polluting factories and noisy highways are negative.

As to assessing impact in terms of features linked to global warming, the subjects that need to be studied also vary a great deal according to the scale. For instance, on a local and regional scale *a higher sea level* is going to be of considerable significance. On the other hand, when looking at the whole world – as in this book – the areas that will be flooded with higher sea levels become so small on a global map that they are not relevant for that scale of planning.

This section *presents maps that show:* 1) *constraints* in terms of future use of land on the globe and 2) *areas positive for habitation* on the globe in the future. Map 3) shows what *areas will gain topologically* because of the new, proposed corridors through continents, both in the northern part of North America and across northern Asia. These northern areas are now opening up for new transportation corridors for roads and railways because of decreased snow cover and a more benign climate. This map is shown on page 88 and presents a development belt around the *linear centre* of these continental transportation corridors. The growing importance of centrality of areas today in terms of development plays a role in making *the ribbons of habitation* along these linear centres important areas in the future. Finally map 4) presents an evaluation of what areas will gain and lose *as the main global shipping shifts to the North,* to the Arctic Ocean.

All these maps are to a large degree based on the fact that *the northern hemisphere* – especially its northern part – will in the future experience more warming and become freer of the present constraints of cold weather than most other areas. The main consideration – as has been much studied in the first part of the book – is the introduction of new major global shipping lanes through the Arctic Ocean, thus connecting the Pacific and the Atlantic Oceans.

The warming is also the main reason why the huge rivers of Northern Canada and Siberia are already becoming more and more ice free. They, therefore, will gradually become very important transportation routes. This means that the areas around them – and elsewhere along the Arctic rim – are going to be very important for future exploitation of resources and habitation. The whole length of the Siberian coastline will become accessible for shipping and resource

Even if it is hard to believe, many cold areas will become very hostipable in the near future.

This section presents maps that show:

1) Constraints in terms of future use of land on the globe

2) Areas positive for habitation on the globe in the future

3) Maps that show what areas will gain topologically because of the new, proposed corridors through continents

4) An evaluation of what areas will gain and lose as the main global shipping shifts to the North, to the Arctic Ocean

exploitation – the most important being oil and gas, as well as timber and minerals. In addition, the Arctic Ocean will also become a very important fishing area for the coastal settlements of the Arctic rim.

The site opposite to Siberia on the Arctic Ocean; the islands of the Canadian Archipelago and Greenland, on the other hand, will not experience as fast a retreat of the pack ice. This will mean that they will not open up as fast to development and shipping as the Siberian coast – especially its western coast – and therefore, in the time span this chapter deals with, or about 100 years, will not be of particular relevance until the last few decades of that period.

Let us now start to study how the first map (see below) – that shows *the bad areas of the globe in the future,* was conceived and designed. First, we have the grey areas; the cold areas earlier mentioned, areas that will be the last to become ice free. Secondly, the map shows rather extensive brown areas, which are mountain areas over 200 metres in elevation. Areas above this height in the high North are rarely fit for habitation because of increased cold and wind and therefore fall in the category of constraints on this map. Even areas between 100 or 200 metres are much more negative in terms of

Ice is retreating from coasts in the Arctic, opening them for transportation and fishing.

Deserts are expanding. With increased warming and less precipitation nothing can stop them, and they will engulf many cities.

activities and habitation in the high North than in more southern areas on the globe.

The most eye-catching areas on the map, however, are *the yellow areas*. The light yellow colour shows today's deserts and the dark yellow shows the areas where the deserts are predicted to expand to with global warming. The deserts of today have very little habitation, so they are not of particular interest for our present studies of feasibility or suitability. The *areas of the greatest concern* are the *areas that the deserts are going to expand over*. These often highly populated areas of today will

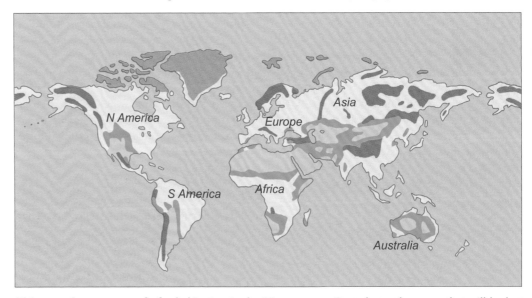

This map shows areas unfit for habitation in the 21st century. Grey shows the areas that will be last to become ice free; brown, mountain areas over 200 metres in elevation. Light yellow shows today's deserts and dark yellow shows the areas where the deserts are predicted to expand to.

The high northern areas that are now getting warmer – and also have sufficient precipitation – are very fast becoming good agricultural lands.

negative way, but not enough to make habitation impossible. *The dark green colour,* on the other hand, shows areas that will be improved by global warming. The areas that will be improved are mostly in the northern regions of the northern hemisphere, but also smaller areas in South America, South Africa, Australia, and New Zealand. In Europe areas that will gain from global warming are all located north of the Alps, though of course, crops and some other types of industries located there will have to be changed or adapted to better fit the climate changes of the future.

In Siberia, it is mostly the corridors along the largest rivers that are shown dark green, and also in Canada, especially in areas close to the Atlantic coastline of Labrador, Newfoundland, and areas by Hudson Bay. In Alaska the dark green areas lie around the Yukon River, and in NW Canada by the Mackenzie River and also along the northern coastline of the North American continent.

The third map that we are going to study *is on the next page.* This map shows areas that will gain in importance because of a better climate and their centrality within a larger area. This map shows two linear centres, in both the North Asian and North American continents. The more northern

in the future be overblown by sand and will experience excessive heat and extreme dryness. There will probably be no way to avoid evacuating people from these areas in the future. The areas that are left beige on this global map are the areas that will be free from these major constraints of global habitation in the future.

The map below has the task of assessing which will be *the most important areas for future development* within the beige habitable spaces defined by the constraints map to the left. The light green areas will be somewhat affected by global warming in a

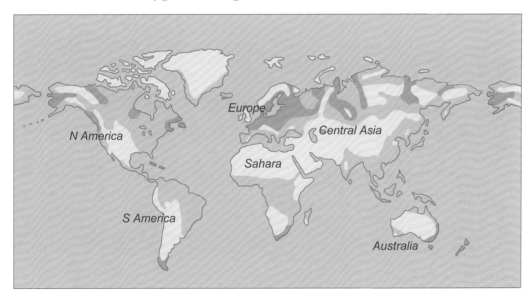

This map shows in green the areas best fit for habitation in the 21st century. The light green areas will be somewhat affected by the climate changes but will be able to adapt and adjust. Dark green shows areas that will become more habitable. Beige shows areas that are mostly uninhabitable.

linear centres are about 1000 km south of the Arctic rim. In North America *the more northern linear centre* connects with the Hudson Bay in the East, to Juneau, Alaska, in the West.

In Eurasia there are also two linear centres shown on the map. The southern one is the traditional linear centre, with China and the Yellow Sea at its eastern end. *The other linear centre* – somewhat to the North – connects the north-eastern corner of Russia at the Sea of Okhotsk that opens into the Pacific, *through the whole length of Siberia* to the Kola Peninsula in its far West. This horizontal corridor will become very important in the future.

The last map of this section, on the right page, shows an assessment of *what areas in the world will gain or lose most* from the opening of major shipping routes through the Arctic Ocean. As these new routes have become highly frequented other traditional shipping routes will lose relatively and, in the process, the areas, ports and countries that now enjoy the benefits of being in contact with these main streams of activity on the globe. The areas that will lose the most are the central regions of the world, mostly the Suez Canal area in Egypt and the Panama Canal area in Central America. Today these canals connect the world oceans, and thus the

continents. (See discussion on page 52 on how much these canals can be possibly expanded).

In the later decades of the century, Arctic shipping routes will start to open through the Canadian Archipelago. These Canadian routes are not shown on this map – nor is their impact assessed – because we, in this map and this section, are only accessing what is of primary importance in this century.

Assessment of the *positive features connected to the new shipping routes* is shown in *three shades of green*. Falling into the *first category* is Iceland because it sits in the middle of the opening of the Arctic Ocean into the Atlantic. Northern Norway will in the next decades be conveniently situated where most Arctic shipping is going to pass.

The areas, however, that will gain the most are on both sides of the narrow Bering Strait where the Arctic Ocean opens into the North Pacific. On the Russian side of the Bering Strait we have Kamchatka Peninsula and on the eastern side of the strait is Cape Prince of Wales in Alaska. The whole western seaboard of Alaska that faces the strait will also be of much importance, not least because there the Yukon River spills into the ocean. In addition, the islands that stretch east from the Alaskan Peninsula and continue as the

Trains are the best trans–continental transportation mode, especially for heavy goods.

Northern bays and waterways are becoming more ice free, opening these areas for development.

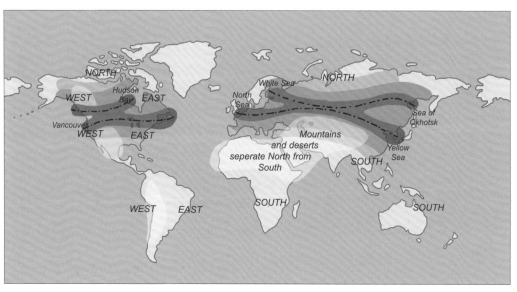

This map shows areas that will gain in importance because of a better climate and their centrality within the better inhabitable North. The two red lines going through North America and Central Asia represent the location of two trans-continental railway lanes, the more northern ones being new.

crescent of the Aleutian Islands will be very beneficially located.

In the *second category* of positive impact, other coasts close to the shipping lanes that will be of importance include south-eastern Greenland, Norway's west coast, the UK, and north-western Europe, in addition to the whole length of the Siberian coastline from the Kola Peninsula to the Bering Strait.

Several things *will make the Arctic sea routes so important*. First of all, are the enormous oil fields that are now starting operation in northern Norway, in the Barents Sea and in western Siberia. This area has now become largely ice free most of the year and transportation of oil in huge tankers has already started, mostly to the east coast of Northern America and to Europe.

The development of shipping will result in *even larger container ships and tankers* than those in use today. For some time now tankers and the biggest container ships have not been able to traverse the Suez and Panama Canals and have had to take long detours south of Africa and of South America. For these tankers, especially, which are too broad abeam to negotiate the bottlenecks of the Panama and Suez Canals, the Arctic sea routes will mean much shorter travel distances between

the Pacific and the Atlantic. The lower part of the map shows today's most frequented global shipping lanes and includes an assessment of which areas will lose most after the opening of the Arctic shipping lanes.

The evaluation of *what areas will lose the most* are divided into *three shades of blue* – dark blue showing areas that will lose most. The areas close to the two canals and the Strait of Gibraltar are thus coloured dark blue. These canals and areas will lose importance, not least compared to the earlier times of warfare. The next category shows that the whole coastline of the Mediterranean will lose importance.

As shipping lanes pass from the Pacific through the Panama Canal they go into the Gulf of Mexico and to eastern and southern America. As ships go the other way west through the Panama Canal they spread to the north to Mexico and California or south to Columbia, Ecuador, Peru, and eventually Chile. Relatively speaking these areas, as the map shows, will all lose importance in terms of global shipping.

One of the main findings of the four maps of this section is that there are corridors across the northern part of the North American and North Asian continents that will gain greatly in impor-

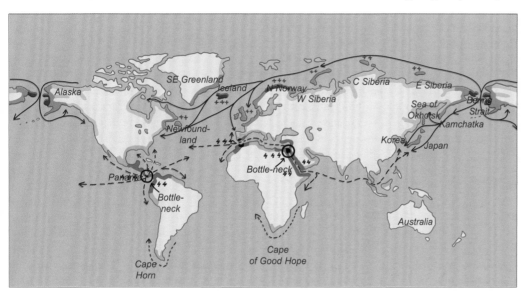

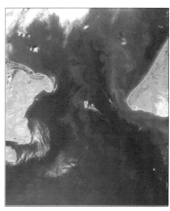

This is an assessment of which areas in the world will gain or lose most from the opening of major shipping routes through the Arctic Ocean. Areas that will gain are shown with three shades of green, and areas that will lose are shown in blue. The findings in short: The North gains, the South loses.

A satellite photo showing the most important watergate of the future: The Bering Strait.

The large Mackenzie River that flows from Canada into the Arctic Ocean will soon become a very important waterway.

The name of the southern tip of Africa, Cape of Good Hope, expresses the feeling of seafarers.

tance. This is both because of better climate and easier coast-to-coast rail and road transportation and because new shipping lanes will open into these now isolated regions as the large rivers in Siberia and Northern Canada become ice free. These north-flowing rivers *will connect vertically* to the *new land transportation corridors,* making the nodes where the continental horizontal linear centre meets these vertical transportation axes very important.

Many areas in North America – primarily the desert areas of Arizona and the Rocky Mountains – will lose but the transportation corridor that goes approximately along the Canadian-USA border will grow in importance. *Another corridor will develop further north* connecting Hudson Bay – that reaches almost into the centre of Northern Canada. This corridor will connect the north-eastern Atlantic coast of Canada to the Pacific Ocean in the West, perhaps at the harbour of Juneau, some 1000 or 2000 km south in the state of Alaska.

The areas north of this corridor in Northern Canada – and also the Canadian Archipelago – will develop later than the northern part of the Siberian coast because the pack ice will retreat more slowly from there than along the Siberian coast. An exception to this is the northern coast-line of Alaska where there are already large oil fields in operation.

Today the oil is transported by a pipeline that goes through the whole of Alaska down to Bligh Reef, where the *Exxon Valdez accident* happened in 1988. Alaska will be very important to the USA because the USA is a rather southern country and will therefore be hit by many problems in the future; thus the 1000 km-long Alexander Archipelago that runs south, from the Canadian coast, will be very important for the USA in the future.

As shown on page 46 the area around the July 20°C isotherm line goes through the Alps in Europe. This shows us clearly that currently the areas north of the Alps are not warm areas except for a few days or weeks in the summer. This position of this isotherm line will not change substantially, so that some warming will not mean pressure on northern and western Europeans to relocate because of a warmer and dryer climate.

Given the relative affluence of these areas the population will also, most likely, have the economic means to buy equipment to help them deal with greater warming.

The areas that will gain most by global warming in Europe are the southern part of Iceland, Sweden, Finland and the area east of Finland in Russia all the way to the Ural Mountains. Areas with a similar climatic status will also be found in central and eastern Asia, but there these areas are mostly the areas of lowest elevation along the large rivers.

As, we in this study, focus on the year 2100 the northern most parts of Canada will have to wait until it gets a higher rating – perhaps another 50 to 100 years. In addition, probably by that time large parts of the Greenland ice sheet will have disappeared, revealing that Greenland has a huge inland sea. What will appear from beneath the ice in the archipelago and in Greenland in terms of land and resources nobody knows for certain: It is quite possible that these areas are rich in resources. Exploration for oil, for instance, along Greenland's west coast has in 2006 led to the estimation that the area has as much oil as was under the North Sea.

Now at last, a few words about the southern tips of South America, Africa, and Australia that in the future will become more benign in terms of climate and habitation.

Note that Antarctica is not shown at all on this or most other maps in this book. This is not because it has no importance in the future, but rather because its release from pack ice and glaciers is so far in the future that it would complicate this study to consider it. There are, however, areas in Antarctica that are already freer of ice, like the Antarctic Peninsula, which is only 1000 km from the southern tip of South America, so there will be surely be certain sporadic settlements there. Argentina has for some time had a "town" in Antarctica. One of the main factors that limit consideration of Antarctica as a site of settlement is the several existing international treaties that guarantee non-development and conservation of resources in the area.

3 Megapatterns of Change

In the first two sections of the present chapter on *The Future Structure of the Globe,* we have studied two basic elements that we need to have some idea about in our quest to understand the future structure of the globe. The first section discussed today's global transportation systems – and then presented ideas on how they might evolve in the future. The second section presented ideas on what will be the good and bad areas for habitation on the globe in the future, taking the impact of global warming into consideration.

The theoretical basis that is presented in the first half of this book introduced the idea of *basic structural patterns* of settlement and the concept of *megapatterns,* a very generalized pattern that deals with large areas and large time frames. The word pattern also indicates that these spatial patterns – that become apparent in historical studies – actually are also patterns for future development. The name of this present section is therefore: *Megapatterns of Change,* with the word "change" underlining the fact that the dynamism of settlement processes are today speeding up because of global warming. Such dynamic patterns are therefore of much more interest in planning studies today than they were earlier.

Let us now introduce the two main categories of megapatterns: First, we have *megapatterns driven by global warming,* and secondly, *megapatterns driven by the governing transportation modes* of each period of time.

This section will make use of these megapatterns in clarifying how certain regions of the globe have developed historically and outline in what way they are likely to develop in the future. This will help us make better use of the features connected to the megapatterns that are *driven by the transportation technology* of a given period. Thus, we start by studying urban developments in the first phases of *industrialization* and *urbanization.* In Europe we see that modern urban development started along coasts and rivers in the 19th century for the simple reason that ships and barges were, at first, the only means available to transport heavy loads of coal, iron and other goods. Later, with railways, roads and waterways, the interior started to develop – eventually, in the 20th century, leading to the fact that the coastal areas of Europe did not experience as much growth as the central areas of the USA where the development of large inland cities took off. The growth of Europe's interior spaces is still in progress, whereas many of the old coastal towns are still in decline.

Let us now proceed in our study by investigating how these *spatial trends* have developed in the last decades. Today a spread to the more comfortable warm, southern areas enters the picture

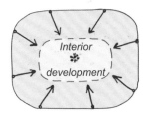

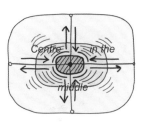

The three most important patterns in the development of modern countries:

1) People migrate to coast, into the new industrial cities as shipping develops.

2) People return to the interior as heavy industry declines.

3) The main centres of the future develop at point centres or linear centres.

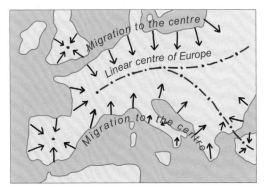

At the beginning of industrialization ships were the only means for heavy transportation, so most industrial cities developed at coasts or rivers.

As interior transportation developed people started to migrate back to the interior. The turn to "light weight" industries helped in this.

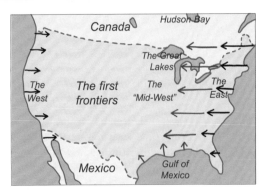

The arrows show how the USA was settled from the coasts. As the coasts had been settled, later immigrants and others moved inland.

An interior core developed around The Great Lakes that provided good transportation facilities. The area was also rich in coal.

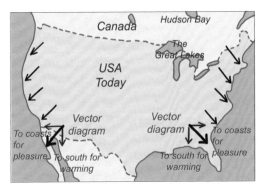

After the prime time of agriculture and heavy industry, and as society became more affluent, a pull to coasts and to the warm south started.

where – in the warmest areas – the migration goes towards coasts.

A brief review of how the USA developed spatially compared to Europe is informative. The USA was at first largely settled by people crossing the Atlantic from Europe, meaning that it was the East Coast that first was settled by people of European descent.

Gradually the coastal areas and the regions closest to the coasts were fully claimed from the original Indian tribes and pioneers started to venture into the interior. Huge waves of further immigrants from Europe, for instance from Scandinavia and Germany, created an impetus to settle the unclaimed lands west of the Appalachian Mountains.

The Louisiana Purchase, acquired from Napoleon in 1803, provided a huge space unsettled by white men. A further need was the search for fertile soil that had not seen the plough and a huge impetus was the discovery of gold in 1849 in California, an area that had already been settled by the Spanish. The editor Horace Greeley, objecting to the negative effects of industrialization in the East, advised young men looking to build their future to: *Go west!*

Settlement of the western part of the present US received a boost with the completion of construction of *the first transcontinental railway in 1869.* In 1914 the West Coast acquired a new transportation connection with the East Coast and the Old World with the opening of the *Panama Canal.* By the late 19th century the Midwest had already been claimed and huge deposits of coal and iron needed for industrial production were discovered. The areas *around the Great Lakes* were premium ground for industrial development, which led to the founding of important cities like Chicago, Detroit and Milwaukee. Across the Mississippi beckoned fertile soil and vast plains where buffalo roamed until overhunted, where cattle could graze, and later where maize and wheat could grow in abundance.

As heavy industry started to decline in this area in the late 20th century people had more means and more time for living a lifestyle that offered them more comfort than the cold winters and the hot summers of the Midwest. Many Midwesterners resettled in Southern California, Yankees left New England winters to enjoy sunshine in Florida, and others discovered the almost constant sun of Arizona and New Mexico. The migration of people from the central areas of the USA south and to warm coasts is still continuing.

With further warming of the climate it is likely that this drift to the coasts will continue, but not so much to the south as earlier. One of the great benefits of living close to coasts is that people have a gradient of climate to choose from, i.e. people can go to the coast or vacation in the mountains on warm days or stay inland on cool days.

It is now predicted that in the near future San Francisco will have the climate of San Diego, which is some 1000 km to the south at the Mexican border. When this happens, it is likely that the vectors on the top map to the right, which show where the mainstream of migration are starting to go, will have started to point northwards along coasts instead of southwards, as shown on the map in the middle.

Another spatial pattern that has more to do with development of inland transportation and the growing importance of air travel is *the growing importance of centrally located areas*. Until 20 to 30 years ago, most head offices of companies and centres of transportation were located in the eastern part of the USA. This was for instance the case when UPS (United Parcel Service) began, with its main central distribution at the Newark, New Jersey, airport. Simple calculations showed that a more central location for its distribution centre in the USA would reduce the total of transportation distances within the USA and so the UPS moved its central operations to Louisville, Kentucky.

Because of this *centrality principle* it is likely that the decline of the middle section of the USA will stop in the future, and possibly even be turned around. It is also likely that the southward drift from the Midwest will be reversed, leading again to an increase in population in that area because of its centrality and its northern placement with a warming climate.

Now as earlier, the Great Lakes are of considerable importance for transportation. The close-

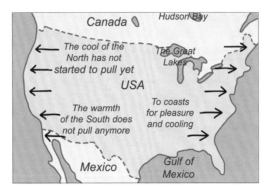

In the first decades of the 21st century the pull towards cool coasts will intesify, but the southward trend will mostly ease off.

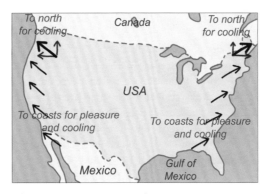

The migration arrows to the coasts start to point north around the middle of the century because of increased heat.

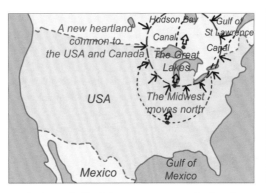

In the later half of the 21st century the border of Canada and the USA has opened and the core of the Midwest has started to move north.

The climate of the Great Lakes region will become warmer, and thus gain in importance.

ness to their cool coasts will also be of growing importance with the warming of the climate. The border between Canada and the USA runs in part through the middle of the Great Lakes and it would be sensible for the development of the whole of North America if the USA and Canada were to negotiate further joint co-operation in responding to the mutual problems that global warming will bring. If that happens, it will help the Great Lakes area to develop more towards the North, which is a sensible direction to go for expansion, given the prospect of further warming of the climate. Analogous to the transcontinental railway system in the USA, southern Canada also has a transcontinental railway system that includes a terminal in Churchill in the north for the exportation of wheat to Europe. Rail connections stretching south into the heart of the continent and as far north as Churchill can facilitate future development.

On the map on page 87 that shows what areas will become habitable with global warming, we see that the largest of these areas are located in Central Asia and Siberia. The quality of these areas for habitation will be quite good as they emerge from under snow and ice – and they will provide ample space for large numbers of people that possibly might want, or need, to migrate there from more southern regions.

It is therefore of great interest to try to form ideas as to *how habitation and urban structures will evolve in Siberia in the future* (see a series of diagrams on the other page). These huge areas are, however, not only of interest because they will become warm enough for various types of agriculture, transportation, and settlement, but also because – as they emerge from beneath the snow – they have clean soil and water. It is likely that the people living in these areas today are not very interested in too many newcomers moving in, even though population is sparse today.

One can therefore assume that something similar will happen as with the settlement of the Americas earlier, that the indigenous peoples will be opposed to an influx of people fleeing worsening conditions where they had been living and intruding on their own home lands. *Such a migration must be planned* with the original inhabitants in a way that they benefit from the influx.

It is very likely that the further settlement of Central Asia and Siberia will happen simultaneously along two paths: Along the coast and along a new northern transportation corridor that will go through the whole length of Central Asia, in a more northern location than the Siberian railway does today.

The text and the maps on pages 71 and 72 have already described the steps to develop shipping along the Siberian coast and in the Arctic Ocean. The shipping lane close to the western Siberian coast is already more or less ice free during the summer months because of the retreat of the polar ice. The problem with this route is that it goes through narrow straits between the coast and islands so that large ships are not able pass through any further than the island of Novaya Zemlya near Western Siberia. At the western end of Siberia the ice has been retreating faster than at its eastern end because the warm *Gulf Stream* current has, in the West, relatively free access into the Arctic Ocean.

Both in the Barents Sea area and in West Siberia *oil and gas extraction* has started in a big way because of this and it is predicted that around 2015 production will be 50 million tons. This is the amount of oil that 500 tankers of 100 thousand tons each can transport.

Even though the shipping lane closest to the Central and East Siberian Coast Siberian coast does not easily allow ships over 10 thousand tons to pass, it is efficient enough to make the Siberian coastline accessible in the near future. This will very much speed up all activity along the Siberian coast so that new industrial sites and towns will be built or earlier small facilities increase in size.

Less frost will also mean that the large rivers that flow north through Siberia and into the Arctic Ocean will become more and more ice free. These rivers will then slowly start to acquire a ribbon of activity along their banks. The other main catalyst for development in Siberia is the new transportation corridor that is being developed for connecting China with the Atlantic Ocean at the port of Narvik in Norway. From there a sea link will help the Chinese to transport their goods to Europe and the eastern sea coast

The text and the maps on pages 71 and 72 have already described the steps to develop shipping along the Siberian coast and in the Arctic Ocean. The shipping lane close to the Siberian coast is already becoming more ice free during the late summer months because of the retreat of the polar ice.

of North America. The harbour of Narvik is located very far north in Norway, a harbour that earlier was closed several months of the year because of ice but is now mostly ice free the whole year.

As in the early phases of industrialization in Europe and the USA, the waterways along the large ship-going rivers are going to become very important for the development of the interior of Siberia. One reason for the importance of these water-ways in the near future is *the thawing of the per-mafrost* which reduces the carrying capacity of the roads and railways based on it. This will mean that the waterways, once they are ice free, will be high-ly important as a transportation system in Northern Siberia. Therefore – as in Europe and the USA earlier – ports in the estuaries of large rivers, where the waterways of inland and coastal waters meet, will be very important as part of a transportation net and will become future sites for urban and industrial development. Eventually water channels will extend the natural waterways further south, into central Asia and eventually, in some places, connecting to the trans-Siberian rail-ways. As elsewhere, such points of connection will become very important nodes in the main transportation system of the Siberian heartland in the future.

As always happens, new transportation possi-bilities drive people to search for opportunities to use these new possibilities to export and import various types of products. Therefore agricultural and industrial production along these transporta-tion corridors will grow, and the search for min-erals, oil, and gas will intensify.

It is a fact that the estimation of available resources is mostly confined to the well-explored regions. Once new areas open for exploration, however, people start to find resources that they could not, realistically, have previously included in their projections of available world reserves.

Today, the general feeling is that the world is running out of many important resources within a relatively short period of time. Yet, as we look at the world map we realize how large the new areas are that are being opened up with global warming. We can therefore reasonably assume that in these vast territories, huge resources, gas

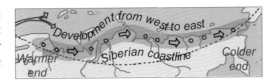

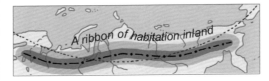

The development that has already started along the West Siberian coast, will continue.

Three steps in the develop-ment of the Siberian coast:

1) The coastline itself develops first – from west to east.

2) Settlement moves inland as the ice on the rivers retreats.

3) A linear center develops inland, along the coast.

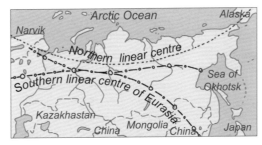

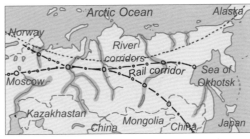

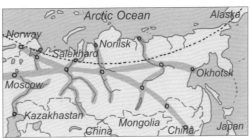

Global warming will mean that Central Asia will become a major settlement area in the future.

Three steps in the develop-ment of Central Asia:

1) A new northern railway line across Central Asia develops.

2) As the rivers have become ice free, they become corridors.

3) At the intersections of rivers and railways cities develop.

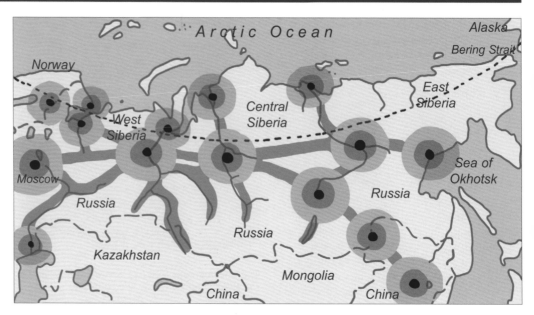

A possible settlement structure of northern and central Asia in the future. Cities are shown at intersections of transportation corridors and waterways shown with broad gray lines on the map. Around the cities (the red dots), are spheres of decreasing influence.

and oil and other resources will be discovered there in the near future. The key is that these areas are becoming approachable because of the reduction of pack ice and snow, which in turn will lead to increased shipping and to the development of other transportation facilities.

We now live in a global world so a globe is necessary in every home to see it as it is.

On pages 8 to 11 there was a discussion of the problems associated by presenting the globe on a flat map, leadingto distortion of shapes and sizes. This map, by the author, is composed of 60 triangle poly hedron, and presents shapes and sizes almost as they actually are on the globe itself.

4 Spatial Structure of the Future

In this fourth and last section on the future structure of the globe, we return to the large spatial systems – of the past and the future – as discussed in chapter 3; *How Spatial Systems Change* (page 59).

At the beginning of this section, it is useful to recapitulate the main findings of the three earlier sections of the present chapter. The first section dealt with global transportation structures. Here two main things will happen. *First:* A new premium global shipping route will open, connecting continents over the North Pole area (see below, and in further detail on page 71-72). *Secondly,* – together with the warming of the northern areas of the globe – this will mean that the linear centre around the globe that now goes through the Mediterranean will, in the late 21st century, have moved north to the latitude of London and Berlin (see page 61).

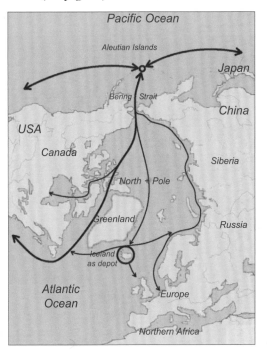

The Arctic shipping routes are starting to open up. By the end of the century several routes will be connecting the Atlantic and the Pacific.

This spread to the north will mean that new, more northern east-west transportation corridors will develop, in both Central Canada and Central Asia – where new, more northern transcontinental railways, will play the biggest part (see page 83).

The second section of this chapter pointed out the good and bad areas for habitation on the globe in about 100 years. The main findings were that the belt of deserts and mountains – which goes from the Sahara in Northern Africa to the Asian subcontinent and then over the Himalayas to China – will widen through the extension of the deserts in these areas (see page 86). This will make the separation between the hot South and the cool North still more definite than it is today in Eurasia.

A similar barrier-belt of deserts and mountain ranges in the Americas, that also will become wider with the expansion of deserts, on the other hand, runs south-north, and is therefore not a hinderness to the south-north migration that is necessary because of global warming. Because of the widening of third belt because of the extension of the deserts that run along it in many places, this belt will still further divide the western areas of the Americas from the more spacious eastern regions than it does today.

The third section of this present chapter on *Megapatterns of Change* gave an overview of several climatic and spatial megapatterns that will influence what the spatial structure of the future will be (see page 91). The two main climatic megapatterns are the *spreading towards the Polar areas* (the first picture in the column to left) and *migration to coasts in the warm areas of the globe* (the second picture). The third picture in the column shows an important megapattern of the future that does not necessarily originate in global warming. This is the trend that *central areas will grow because of improved land and air transportation.* If, in addition to improved transportation, some of the areas are emerging from the bondage of snow and cold – like Siberia and Northern Canada –

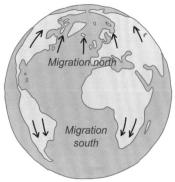

Megapattern by global warming: Towards the poles. A basic feature in the structure of the future.

Megapattern: To coasts in warm countries, which means reduced importance of their interior.

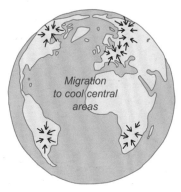

Megapattern: Central areas will grow if they are not too hot, i.e. primarily in "cooler" areas.

The stars of the EU sign represent the twelve first member states of the union.

this will contribute greatly to the growth of these central areas.

The assessment of all these things connected to the warming of the globe can be summarized in *two main findings*: 1) The centre of Earth will be much worse off, and 2) the Polar and sub-Polar regions will – in most instances – be much more habitable.

In the first half of the book, four spatial features of the globe in the future were outlined theoretically. This chapter reviews them and gives examples that provide an idea about the spatial structure of the future.

First however, let us start by reviewing the development of the spatial structure of Europe in the last 50 years. The *European Union* (EU) was established in 1957, composed of five central European countries and Italy. Sixteen years later, in 1973, its extension to the north started, as Ireland, the UK, and Denmark became members. The extension to the south occurred with the admission of Greece in 1981, and Portugal and Spain in 1986. A step toward expansion to the east came with Austrian membership in 1995. The same year saw an important event in the expansion to the north as Sweden and Finland

entered the union.

Further extension to the north has not yet been successful. The Norwegians have twice said no to EU membership in a national referendum. In place of membership came a special agreement between the EU and Iceland and Norway in 1994, called the *European Economic Agreement* (EEA). The main reason why Iceland, Norway, Greenland, and the Faroe Islands have not been willing to enter the union is the stubborn adherence of the EU to the rule that all territorial waters should be common to all EU member states. That rule is simply unacceptable to these northern countries that base their livelihood to a large degree on marine fishing and thus also on their own control of their fishing grounds and marine resources.

On May 1st, 2004, the next big step in the extension of the EU happened as ten eastern European countries entered the Union, (Estonia, Latvia, Lithuania, Poland, the Czech Republic, the Slovak Republic, Hungary, Slovenia, Croatia, and Bosnia-Herzegovina).

The EU is now preparing a further extension of the union to the east to include Belarus, Ukraine,

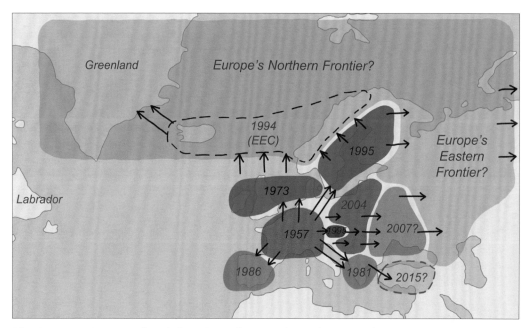

The European Union was founded in 1957. The map shows the years of its extension. The eastern additions remove a barrier from connecting with Russia, which is hugely important because of its Siberian resources. A northern frontier would include Greenland, Iceland and Norway.

Moldavia, Rumania, Bulgaria, and also Turkey. These countries are not strong economically nor do they have many proven natural resources. They, therefore, probably will continue to be rather problematic areas, not least with increased warming of the climate because they rely so much on agriculture. These new additions will therefore probably prove to be more of a burden, rather than strengthening the Union.

Of course, the central European countries will acquire new markets in these new, largely under-developed countries, factories can be relocated to make use of the labour force at prices competitive with the third world. But the factors involved in these newer admissions need to be considered in terms of an extension policy for the EU, as the future is not in the East nor the South, as this book describes, but in the North and the North-east as seen from Europe.

A wise policy of the EU would be to relinquish its claim on the territorial waters of the northern European countries and to do everything *to open up a Northern Frontier for the Union,* where there is ample space and enormous new resources for future generations. Oil, for instance, has been discovered in Greenland and the oil fields of Norway stretch further north towards the Barents Sea where Norway abuts the Russian border. European Union needs to include Belarus and Ukraine so that they are not a barrier as it comes to a possible future extension of the EU to Russia. Russia is already of enormous importance for Western Europe because of its resources and its huge northern areas that have a large potential

for future development in a warmer world.

Before the fall of the Berlin Wall in 1989, Berlin was outside of Western Europe. From a different and more valid future perspective, Berlin will almost be at the centre of a large circle that embraces most of Western, Eastern and Northern Europe. The next big circle of this type is Moscow. Still further east another major Eurasian centre – closer to the centre of Asia – might develop at the Ural Mountains (see below).

As has already been pointed out, *the geographical centre of the land areas of the globe* is close to the southern end of the Ural Mountains. With examples earlier in the book we have seen how the geo-

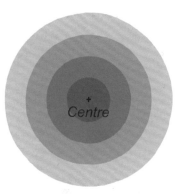

The power of the centre of a circle, means a strong tendency of functions to move to there.

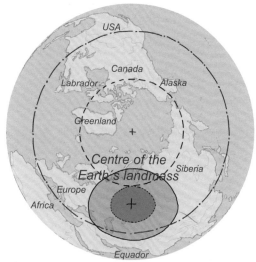

The centre of the landmass of Earth is at the southern end of the Urals. Global warming and its location will empower it in the future.

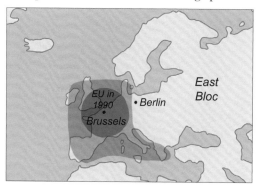

Until the collapse of the East Bloc in 1990 Berlin was outside the EU. The right figure shows how much the picture can change in a short period.

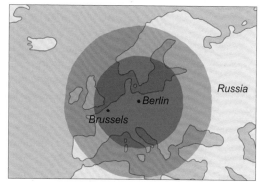

With the extension of the EU to east and north, Berlin becomes a logical central point as the capital of the union.

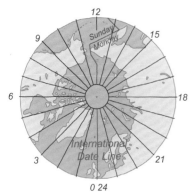

In the extreme North all time zones become one. This calls for one time for the whole Earth.

Seen from the Arctic, vertical connections are more important than today's horizontal ones.

Inside buildings an environment of daylight and good temperatures can be created in the Arctic.

graphical centres of regions are, and will be, of growing importance in the settlement structures in the future. Today of course, it looks somewhat strange to suggest that this area in the Urals could possibly develop into an important central area in the settlement structure of the whole globe – especially since today, it is in a very remote location. Nevertheless, *the power of the geometry of the circle* and the quality of its centre factors basic in all settlement development. We can see the power of this spot in cities where the central areas are always of the most importance. The planning of a new development, to take another example, optimally foresees the central area as the site for the most important functions. The necessary precondition for the development of such a Ural centre will be improvement of the climate and improvement of transportation links into and through this region. This development will take considerable time, but even in this century, we will be witnessing more and more the strengthening of this central area.

Now someone may ask, "What about the claim that was presented earlier in the book, that *the North Pole* would become the spatial centre of the semi-spherical world of the future, with an active Arctic rim around it?" Yes, the North Pole itself will become an important centre in the future, symbolically, even if primarily because of its position as the northernmost point of the globe. Adding to the importance of the Pole is that fact that the climate zones run in successive circular belts south from the North Pole area. Additionally, the Arctic Ocean is approximately circular in shape, with a coastline of some seven thousand km, an enormous coastline that will become easily approachable as the pack ice of the Arctic Ocean disappears in the 21st century.

The North Pole area will also be strengthened by the fact that there are huge landmasses in all directions that will have access to the Arctic coastal waters and the Arctic Ocean shipping lanes. The great rivers that flow through northern Canada and Siberia into the Arctic Ocean will provide easy access to the more southern terrains of the northern hemisphere. Though probably for the next 100 years not many people will actually be living along the Arctic rim, its importance

for resource exploitation – mainly of gas and oil – will be vital, together with very easy access to both the Pacific and Atlantic Oceans, which will make this area economically and politically of key importance in the future.

The Arctic will be a strange area to live in, with its four months of almost complete darkness, two months in the spring and autumn when there is a balance of day and night, and the four summer months where most of the 24 hours are bright with no real night. In the summer, this fascinating world will become more popular with tourists on cruise ships.

Modern technology is making it ever easier to create microclimates within building complexes or glass domes, where daylight and darkness are technologically controlled and all the climate factors are adjusted to the optimum. These kinds of structures will make it easy for people to live in the Arctic during winter, little affected by the total darkness and the high winds and cold weather.

A special positive feature of the four-month long night in winter and four-month long day in summer is that the human biological clock, which must be adjusted to the rhythm of day and night, will not be disturbed by horizontal or vertical global travelling in the Arctic if days and nights are created artificially. In addition, it is of special

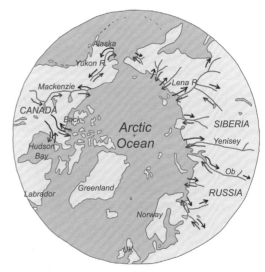

As the Arctic Ocean and its rivers become ice free, North America and Eurasia can interact with world shipping in a way not known before.

advantage, at least for a possible symbolic floating *Capital of the World* at the North Pole, that it belongs, at the same time, to all time zones of the world. This therefore is an ideal location for a financial centre that serves the world, whereas other financial centres such as New York, London or Tokyo, belong to specific time zones which therefore affect calculations of stock prices and timing of bank transfers. As *the spatial system of the North-centred world* has taken over from *the horizontal ribbon of habitation around the globe,* the lines of travel will be more and more from north to south, instead of from east to west. This is much easier for humans than to travel across time zones where the adjustment of one's internal biological clock is disturbed.

It is therefore likely that the areas on the globe that will be working together in the future will not involve as much running along west-east axes as they do today, but rather along the south-north axes pointing toward the North Pole. Some people might point out that going from warm southern areas to cold northern areas will cause problems in transportation, but most railways and airplanes in a warmer world will be able to cope with that. Today most of the continental global transportation structures *go horizontally,* but in the future these structures, for instance railways, will increasingly be *laid out in north-south directions.*

To illustrate this with the case of Europe we today mostly think about the axes of Europe extending from west into east, but other axes that go, say, from Italy through Switzerland to Scandinavia and through Russia into the Kazakhstan, Pakistan, and India, will also gain strength. The Alps in Europe and the Carpathians in Eastern Europe of course present mountain barriers to be conquered, but the digging of more tunnels, already a reality today, can surmount these barriers. Earlier most transportation was constrained to follow and stay within the climatic zones. Earlier ethnic and often linguistic relationships that followed these horizontal belts of the globe were also a hindrance to vertical inter-communication. With a *multiracial future* and the *decreased importance of open-air agricultural industries,* as well as *transportation that is freer from the forces of climate,* relationships and travel across climatic

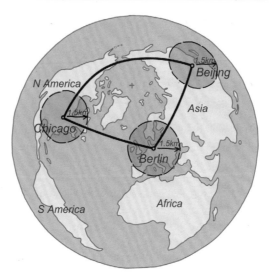

If a new system of high speed trains connecting continents becomes a reality the links to other transportation systems becomes of central interest.

zones – where the vertical axes of development point towards the North Pole and the Arctic Ocean – will increase in importance.

On page 83 an idea of an engineering group about a new transportation technology for the future was presented. This *railway vacuum tube* will connect continents by traversing long distances on the surface, within the crust of the globe, or floating above the bottom of oceans (See the picture to the right). This book has proposed that there should be three main stations in this future high-speed train system, one in each of the three main continents: *Berlin* in Europe, *Chicago* in North America and *Beijing* in South-east Asia. The main technological advantage of such a train is that in a vacuum tube there is very little air resistance against speeding trains and friction is further reduced if the trains run on magnetic cushions.

Another huge advantage is that theses trains could be faster than jets, and also that – like the high-speed train system in Europe today – they would be designed to go directly into main transportation centres, be it major international airports or centres of metropolitan areas, where all types of transportation systems come together. We have many examples in the history of settlement on Earth of how major transportation systems have been decisive in how settlement struc-

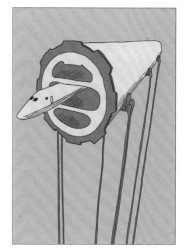

A section through a vacuum tube for high speed trains that would run on magnetic tracks.

Container technology has led to "wholesale" transporation where huge numbers of containers are brought to a "depot" and distributed.

tures have evolved. If the decision is made to select these three major cities as the centres in this future global transportation system, the result will have a very decisive influence on what the spatial structure of the future will be, not only in the regions or continents in question but also over the globe in its entirety.

The system will be very expensive to build and to run, even if the energy required is minimal compared to air travel. It will therefore be too expensive to transport things other than people and highly valuable goods. Bulky and heavy goods will continue to be transported by ships via the world oceans.

As for the structure of future sea transportation, we will also see *a reduction in the number of transportation centres.* The general trend today is that ships are becoming bigger and bigger because increase in size reduces shipping costs. This means that the ships of the future will not be able to enter many of today's shallow harbours, where there is also often not enough space for the vast numbers of shipping containers transported.

The necessary ports require a great deal of land area for unloading and storing cargoes, space that is too expensive in some of the most densely populated regions of the world. Therefore there is *a trend towards huge transshipment ports,* two or three in each main region of the world. From there smaller ships transship goods along a "hub of the wheel" distribution system into specific market areas.

This book suggests two such harbours at each end of the Arctic Sea route, that is, in Iceland and the Aleutian Islands. An important reason for building *a depot harbour* at both ends of this very important global shipping route of the future over the Arctic Ocean, is that the ships going there in the first 100 years will have to be especially built because of the danger of drift ice.

In the earlier periods of the 21st century these huge ships, connecting the two large world oceans, will have to be built with some icebreaking ability, which will make them very expensive. This means that the travel distances that these large, specially built ships need to go, have to be kept at a minimum, making the idea of two depot harbours, are at each end, a feasible idea. If this

idea becomes reality, these specially built ships would only be used for transversing the gradually thinning sea ice area of the Arctic Ocean between these two big depot harbours.

Let us now finally look at *the big picture of the globe,* looking north from Europe, which makes most of the land mass of the world visible in one picture. In this picture, which appears *on the cover of the book,* the main features of the world of the future are presented.

At the top of the picture is the symbolic World Capital at the North Pole (a cross), the highly important Arctic sea routes, and the depot harbours in Iceland and the islands south of the Bering Strait, from where goods can be transported in all directions.

The picture also shows with dark green hatching, the areas of the high North that can be expected to gain the most through the structural changes and the global warming that will occur in the next 100 years. The little brown circles suggest the locations that are most likely to develop into important coastal cities in the future.

The next chapter deals with assessing *the impact of the picture of the future* that has been drafted in this chapter, *on a global scale.* The chapter starts with two sections that summarize what areas will gain and what areas will lose. The findings of these two sections will be used as a foundation as, in the third section, i.e. on *the impact on the future global economy* will be assessed. The chapter closes with a section on *the impact on geopolitics* where, for instance, the location new transportation corridors and new vital resource areas will be of central importance.

The sixth and last chapter of the book is called *The North: The Future Area of the Globe.* There we will get a closer look at the most important area of the globe in the future: *the North,* with a more detailed description of its specific features. The final section describes planning processes that are already being developed to guide and reduce the impact of all of this forseeable future activity on the natural and human environments of the high North.

Reykjavik has already become a depot harbour for some transportation in the Arctic.

5 Impacts on a Global Scale

1 Areas that will Gain

This second half of the book: *The Future*, started with a chapter that outlined what the future structure of the globe will be. The present chapter, which is also divided into four sections, will assess and interpret what this *new picture of the globe in the future* is going to mean in practical terms. The first two sections will describe what areas will *gain* and what areas will *lose* through this development. The third section describes the impact on the *global economy* and then, in the fourth and last section, the impact on *future geopolitics* will be outlined.

It is common knowledge that people from dissimilar backgrounds define differently what is positive or negative, what can be considered a *gain* and what can be considered a *loss*. One of the characteristics of the human condition is *to be afraid of change*, despite the age-old saying that: *nothing is certain but change*. Because of this, most of the changes that come with global warming are judged as negative for example in *Environmental Impact Assessment Reports*.

A great number of reports have been written and published on the impact of global warming. One of them is *The Regional Impacts of Climate Change* published by the *Intergovernmental Panel on Climate Change* in 1998. The description of changes in the Arctic on page 10 states: "Direct effects could include…sea- and river-ice loss…." This wording takes changing environmental conditions in the Arctic as negative. The author of this book is from Iceland, where sea ice has been the worst enemy of the country, which has had to rely on fishing and ocean transport to remain, through the centuries, in contact with the rest of the world. The Icelandic poet Matthias Jochumsson expressed this view with the words: "Are you here again, you ancient devil of this country…." The wording of the report, which *assumes that the loss of sea ice is negative*, therefore comes as a surprise to Icelanders and probably other people living in these coldest areas of the globe. This type of judgement is not an isolated

example. These reports are all more or less imbued with the belief *that climate change* and its impact on the environment *are negative*, even if this means less ice, a better climate and more and different vegetation for Arctic countries.

This wording and the attitude of more southern countries that have basked in a warm climate with easy ice-free transportation indicate that these scientists seem not to be able to put themselves in the position of the people of the high North. This attitude is something that reminds us Icelanders – earlier a colony of Denmark – of times when *the elite in Copenhagen* decided what would be good for us. Every summer we get visitors in Iceland – well educated and caring people – who spend their holidays trying to stop the modernization of Icelandic society, for instance by trying to sabotage the building of hydropower plants. These young people are mostly from the USA and Western Europe, i.e. from areas that have built their wealth on the exploitation of natural resources and the alteration of their natural landscapes to industrial landscapes. These young rich Westerners hardly have the right to express their indignation of a country that wants to develop and to assure its citizens a standard of living comparable to their own. And when these beautiful people come to Africa they say: "Your country

In northern areas the arriving of sea ice means isolation and cooling of the climate. Therefore its reduction, with global warming, is a blessing.

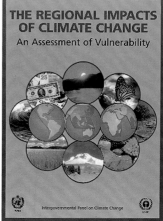

THE REGIONAL IMPACTS OF CLIMATE CHANGE
An Assessment of Vulnerability

Intergovernmental Panel on Climate Change

Most southern reports on global warming interpret reduction of sea ice as negative.

"Global warming"

"A new ice age cometh"

"Slowing of the Gulf Stream"

"Gulf Stream will not reach North"

"Population explosion"

"A meteor hits the Earth"

"Lack-of-copper doom"

"Piles-of-manure in cities"

"The end of oil reserves"

People seem to like doom predictions. The list above shows a list of afew such predictions.

Change: A Disaster?

Change is trouble,
Change is hardship.
But we can rephrase:
Change is challenge!

We can gain from change,
Or, let it defeat us.
The same with global warming:
We can gain from it.

shall be *a global national park,* so we can, as nature lovers, come with our cameras and sometimes our guns and record and occasionally kill your beautiful wild animals." Some people call this *neo-colonialism.*

This book expresses the somewhat different attitude of Northerners: Most of the changes that come with global warming – like higher temperatures, more vegetative growth and less snow – are considered better, even though this may lead to a temporary imbalance in the ecosystem. Many Northerners also consider the improved shipping possibilities – because of the retreating of the sea ice – as positive. It should be easy to understand that people who have lived in relative isolation for hundreds of years welcome more warmth and improved transportation. This book also considers the discovery of new oil oilfields a gain for the local economies and also for the world economy.

Of course, there are different points of view. Some may be of the opinion that a better climate in the North is bad because it will mean changes in the natural environments, even though the changes, overall, will make the North a more productive and vibrant area. Improved transportation – for instance more shipping in the Arctic – will probably also mean to some a negative result because, as in other areas of the world, it will bring a risk of pollution and a loss of tranquillity.

This conservative attitude towards environmental change means that almost all changes in nature, as for instance the retreat of glaciers, snow and ice, are deemed to be negative – no matter whether the effects are going to be positive or negative for the community in question. Simply the fact of change scares some, not least the younger generation.

Even to grow up seems scary to many young people. Many act like youngsters still in their late thirties, and stay in their parents' home until forty or fifty. This book sends the message: *Let us not be afraid of change,* let us embrace the current and impending climate changes as challenging and rewarding opportunity... because this is a change that can't be stopped.

The main reason why change is most often deemed negative is probably the assumption that it is not within our power and technological ability to maintain our present lifestyles. Unfortunately very often *future projections* are simple extrapolations of various negative trends like the depletion of minerals or fossil fuels.

Looking at the *huge victories of science* in the last 200 years that have overcome most problems of our modern times, it appears likely that technological capabilities and advances will be able to solve most of the problems that mankind will

THE
SHOCK
OF
THE NEW

THE HUNDRED-YEAR HISTORY OF MODERN ART·
ITS RISE. ITS DAZZLING ACHIEVEMENT. ITS FALL

ROBERT HUGHES

Modern art helps us get used to change by shocking us with things we have not seen before.

Industrial nations alter their natural environment to build wealth, but want to turn other countries into global preservation parks.

Preservation parks in Africa are in demand by Westerners. Maybe it is also appealing to some to keep the indigenous peoples at a primitive stage.

face in the future. Many people interpret the problems facing the world today *as doom with no solution in sight;* all is black and depressing. In earlier centuries there were *countless predictions of doom;* predictions of the end of civilization. Technological solutions proved the doomsayers wrong. For instance, the "lack-of-copper doom" was solved by the invention of glass fibres, the "doom of piles-of-manure in cities" that relied on horse and carriage transport was solved by developing motor vehicles, and the "doom of the end of oil reserves" was solved, at least for now, by the discovery of new oil reserves, better fuel efficiency and new ways of producing energy. The human condition seems to include *a certain love of doomsayers;* they attract attention and tend to become heroes, even if they are proven wrong time and again.

Another characteristic of the human condition, especially in the West, is to be shameful of man's actions in nature. If a beaver builds a dam it is a wonder, if a man builds a dam it is ugly and evil. This attitude is sometimes called *shame culture.* Shame culture is the idea that humankind should be ashamed of itself – perhaps because of Adam's and Eve's bad conduct in the Garden of Eden that led to their being driven out. The result of this is that it is often the first instinctive response of a Christian to say or think: "What I am doing to nature is a disgrace to the work of God."

This prologue gives *the background to the philosophical standpoint* characteristic of this section and this book. This standpoint postulates that humankind should not be ashamed of its various cultures and its modern ways. This point of view is also based on the approach that there should not be uncompromising proclamations like "global warming is bad" but rather that the impacts should be *judged in context,* i.e. on whether the change is going to be a gain or a loss for the region in question.

In estimating the impact of global changes in this section, we are examining *four time scales*: a) the near future, i.e. the next few decades, b) after about half a century, c) in about one hundred years, and d) after two hundred years, i.e. far into the future. These four time scales are very essential in trying to assess what the impacts will be

and when they will occur. There are, for instance, predictions that the sea level may rise, even 5–6 metres, in perhaps 200 years or more. This is so far in the future that it makes little sense even to consider these numbers. The prediction about the sea level rise is 50–90 centimetres in this century, is however, of practical concern.

There exists a theory that says that the melting of the ice will lead to a lower salt content of ocean currents coming from the Arctic. This speculation – there exist no long-term measurements to support this theory – continues to predict that a lower salt gradient will mean that the cold currents will flow farther south of Greenland and cool the East Coast of North America, as depicted in the film "The Day After". In theory it is also possible that the "conveyor belt" of ocean currents that brings warmth to the North Atlantic will slow down because of the lowering of the salt gradient, which is the main motor that drives the conveyor belt.

This theory does not recognize the fact that the Greenland and Antarctic ice sheets are mostly at such high altitudes that such extensive melting will not occur in the foreseeable future. Furthermore, to confound the issue, there is also an opposing theory that says that the salt content in Arctic waters will increase. The argument to support this is that the longer ice-free periods of the Arctic rivers and the thawing of the permafrost means that hugely increased amounts of sediments and salts are already being carried with the rivers into the Arctic Ocean – and there exist measurements that show this! If, in fact, the salt content increases it might well be that the conveyor belt will speed up, thus carrying more warmth to the north rather than less.

Let us now start to review and summarize *the gains of global warming* in fifteen main regions of the world, starting with the Arctic. Not much will happen in Greenland in this century. Practically speaking, today there is only a limited amount of habitable area at its southern end and very little elsewhere. The slow retreat of the Greenland ice, however, will mean that gradually more land will become ice free. What will be of most gain for Greenland is the recent discovery of enormous oil fields off the western coast, probably contain-

God drove Adam and Eve out of Eden; the main reason for the Western Shame Culture.

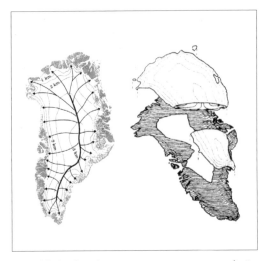

Valuable land and resources may appear as glaciers retreat. The left picture shows that the centre of Greenland beneath the ice is below sea level.

go over the middle of the Arctic Ocean. But the remaining of the sea ice in the middle of the Arctic Ocean in the next few decades, will mean that most shipping lanes will go along Siberia and Norway and from there to Western Europe and North America. Greenland, however, in the further future, could become a very important area. Nobody knows as yet what can be found under Greenland's ice sheet.

Valuable minerals and oil have been found in the west of Greenland and much more could lie under the present ice cap. Ice scopes have mapped the land area beneath the Greenland glacier, revealing an inland area that is below sea level that has a little opening into the Davis Strait to the west as the picture to the left shows.

Iceland and *Norway* are those Western European countries that will gain most from the opening of shipping routes through the Arctic Ocean between the Atlantic and the Pacific. These routes are already opening in the summer further and further to the east along the Siberian coast. In only a few decades they will be readily used in the summer months and much longer with specially strengthened ships. What, however, is of most importance for the near future is the development of oil production in the *Barents Sea* and the

ing more oil and gas than was discovered under the North Sea. Today the population of Greenland numbers only about 60,000, but with the exploitation of the oil reserves and the warming climate the number of inhabitants on the west coast will likely rise substantially. The east coast and the southern tip will be well located in terms of shipping as the main shipping routes start to

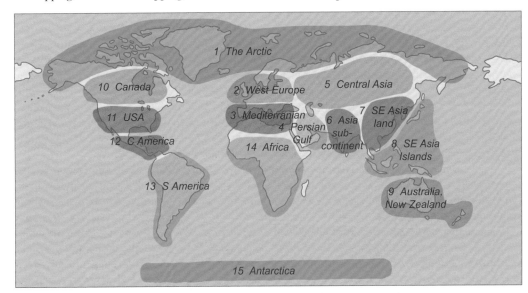

On this map the world is divided into fifteen regions that will experience in a similar way, the gains and losses that result from global warming. Generally speaking, the high North and the far South gain from global warming, but the central hot and arid areas lose.

transport of oil by sea from *Western Siberia*. It is estimated that in 2015 a total of 500 tankers of 100,000 tons each will pass Iceland annually, headed in each direction. This will pose *a grave environmental danger for Iceland* because it is possible for such a tanker to lose power and to drift ashore, destroying spawning grounds and coasts, as happened with the *Exxon Valdez accident* in 1988 in Alaska.

The same danger also looms for the whole north and west Norwegian coast, but the Norwegians have been so long in the business of offshore oil production that they have all the preventive measures in place to meet crisis situations. They can, for instance, deploy extra strong tugboats in a very short time to keep tankers from drifting onto reefs or the shore.

Because of less ice and therefore better access to oil and other natural resources, the colder coasts of Siberia towards the east will gradually start to be developed. The further east along the Siberian coast, the colder it gets, meaning that the eastern coast is less easily accessible to shipping. The same holds true for ice on rivers; it will first disappear in the western part of Siberia.

As these rivers become ice free, they will be the perfect transportation channels for heavy goods to and from the interior of Siberia and Central Asia. This eastward and inland process will continue throughout the 21st century. The narrowness of the Bering Strait, where the Arctic Ocean opens into the Pacific, will mean that there will be enormous shipping traffic there later in the century. The site of major traffic lines always provides possibilities for development, as along the most frequented streets in cities.

Next we will look at the impacts of global warming on *Western Europe*, north of the divide created by the Alps. This is one of the areas on Earth that will gain most from global warming, even if it means changes in the ways people conduct agriculture. The summers and the winters are rather cold in Western Europe, so an increase in temperature is welcomed, especially if it happens in the winter, as now is the case. As the Arctic Sea routes open, *the northern part of Western Europe* with Norway, Great Britain, Denmark, Germany, Holland, and Belgium *will be very well placed in terms of global shipping*. Moreover, Western Europe like the USA, will gain very much from the opening of the oilfields in the Barents Sea. And Europe, already today, is gaining enormously from the Russian gas that is piped into Europe from Siberia.

The third area on our map is the *Mediterranean space*, i.e., Europe south of the Alps and including Northern Africa and the Near East. This area will run into some problems because of increased heat and the rate of evaporation, leading to lack of water and an increase in wild fires. This, however, will not be very serious in the long term because once the forests have been burnt they will be replaced by desert vegetation that is more adapted to a dry climate.

Area four on the map is the *Persian Gulf*. This area is already very warm, but since it is very rich because of oil, most of its inhabitants have the means to cool buildings sufficiently for comfort. A great roofed-over ski area has even been created in Kuwait! Nevertheless, difficulties could arise in say 30–40 years as the oil reserves start to be depleted. If, however, these countries have by then already built up societies capable of creating viable incomes not based on oil, life can be sustained in this area in spite of the increased heat.

The fifth area on the map is *Russia, Siberia and Central Asia*. This rather cold northern area will gain much from global warming. Transportation access will be greatly improved by new transcontinental railways, and as the rivers leading into the interior have become more or less ice-free, this area will be able to reap the benefits of the enormous shipping traffic along the North Siberian border.

The sixth area on the map is the area to the south of west Central Asia, the *Asian Subcontinent*, composed of India, with Pakistan, Bangladesh, and other countries. This will be one of the worst problem areas of the globe, together with *Central Africa*. These are highly populated countries, and there is, in some cases, little space for people to escape to from problem areas like the estuaries in Bangladesh that will become increasingly prone to flooding. In addition, this area is largely closed off from the cold North by the Himalayan Mountain range, ruling out a seamless northward migration.

An indoor winter sport area is even in Kuwait. With cheap energy nations in hot countries can create comfortable micro-climates.

The typical oil tanker of today is 100,000 tons. Tankers of up to 500,000 tons are now being built.

The increased number and intensity of cyclones that come with a warmer climate is going to be a huge problem in the coastal areas of some southern countries.

The area's lack of natural resources and poverty will exacerbate the problems of adjustment, because richness in resources, together with wealth, can help solve so many of the problems connected with global warming. Rich countries can, for example, purify water and the flooding of coastal areas can be averted by levees, as for instance in Holland, if the means to do so are available.

Area seven is *the mainland of South-east Asia*, with Korea, China, Vietnam, and other countries. The southern part of this area will be hard hit with heat and drought. Some relief can be gained by migration to coastal areas, but this will also expose the population to the danger of hurricanes, storm surges and the rise in sea level.

The eighth area is the *South-east Asian islands*, which – because of their oceanic climate – will not be hard hit by the increased temperatures. The most northern of these islands will gain much from the warming. These are the islands north of Japan: the Sakhalin and Kuril Islands, the southernmost ones belonging to Japan and the others to Russia (though claimed by Japan). The whole area around the Sea of Okhotsk and the Kamchatka Peninsula in Russia will also gain very much.

In area nine, *Australia* and *New Zealand*, we will see expansion of the deserts in Australia, but its southern and colder part, where Sidney and Canberra are located, will not be adversely affected. New Zealand will probably experience an overall gain from global warming.

Together with Siberia, *Northern Canada* is the area on the globe that has the largest areas that will become habitable in the future. This, however, will happen somewhat later in Canada, at least later than in Western Siberia. Perhaps the development of Northern Canada will not have taken off till the latter part of this century. The first areas to develop will be Labrador and Newfoundland on the Atlantic coast and, once Hudson Bay has become ice free increased shipping access to the Canadian heartland will be facilitated, together with the development of highly valuable new fishing grounds in the Arctic.

Predictions say that the far north Canadian Archipelago and Greenland will be some of the last areas to be ice free, so even though regular shipping will have started to go through there in the latter part of this century, not much is likely to happen there, unless valuable resources are found, which geologists think is quite possible.

The USA is in part a rather southern country, Florida and the US Mexican Gulf coast, for instance, being at a more southern latitude than the Mediterranean. This area will therefore suffer in some ways from the greater likelihood of damaging hurricanes and the greater summer heat and dryness. At the latitude of the Great Lakes, especially north of the Canadian border, however, there is ample living space that will have a very comfortable climate in the future. It, therefore, would be a very wise move of the USA and Canada to agree on future free movement of people, goods, and investments for the benefit of both countries. And a northward migration into Canada would become easy.

Central America has a long coastline and many islands that enjoy the cooling effects of the ocean. Hurricanes, however, are a problem there as in the USA, and the area will lose some importance as the Panama Canal will become less important as a major traffic corridor for global shipping.

The northern part of South America is located very close to the Equator but once we come further south, the climate becomes more tolerable. The southern part of South America is today a rather cold area, so it will gain from global warming.

Now we finally come to the *Antarctic continent* that is an island in the middle of the southern hemisphere, surrounded by the Southern Ocean. Similar to Greenland, much of the ice there is at high altitudes so it will melt only slowly inland, whereas, on the other hand, coastal ice has already started to break up considerably. There will therefore soon develop coastal settlements in Antarctica in addition to the present Argentinean "town", especially on the tip nearest to South America, again depending on whether valuable resources are found in the area and the present agreements that ban development are abrogated. International agreements have been entered into assuring the conservation of Antarctica, thus constraining development.

Huge blocks of ice have started to break off in the coastal areas of Antarctica.

2 Areas that will Lose

Like the last section on which areas of the globe will gain from global warming, this section on areas that will lose also needs an introduction.

There are four main aspects of future development that will contribute to the worsened situation in several areas of the globe in the future. These four aspects will now be described, along with changes in future development and measures nations can undertake to reduce the negative impacts. The four aspects that can lead to a worsened situation in the future are:

a) *Increased heat in the warmer areas of the world* that already today are at the edge of what can be considered tolerable for people, animals and vegetation – is going to be a great threat in the future. One of the most serious consequences of increased heat is *increased evaporation*. Many of the warm areas of the globe already lack sufficient water, and an increase in evaporation will exacerbate the situation.

Today the price of gas and oil has become very high, so the areas that possess plentiful oil reserves are very rich. The main problem in building economies on oil and gas is that they are not a renewable resource and with the greatly increased consumption of oil in the world the oil fields of these nations will run out one after the other. To put it briefly:

A new gas liquefaction plant in Melkoyja outside Hammerfest in North Norway. Tankers with huge pressure tanks are used to transport the gas.

b) *Areas that run out of oil will lose.* If the situation in terms of finding energy resources, or methods of producing energy in a way that is independent of location, the oil producing areas, more than other areas of the globe, will relatively lose. The biggest hope of mankind is *nuclear fusion* that can use many kinds of materials to fuse and release utilizable energy. On the other hand, depletion of oil reserves will not be serious for those countries that have had the wisdom to invest the profits of their immense oil production in ways that make them independent of today's income from oil. *Coal is an alternative to oil* in the production of electricity but not for powering cars and airplanes. Coal is the main source for energy in for example China and India and their coal resources will last for a very long time.

c) *With the increase in use of the Arctic sea routes, some countries will lose their present advantageous position in global ocean shipping,* mostly in the latter part of the 21st century. This factor has been explained rather extensively on pages 52 and 89.

Fourthly, there is a rather abstract feature of a worsened geographic situation of some areas of the globe, primarily because of:

d) *The reduced importance of today's linear centre and ribbon habitation around the globe.* Because of the warming of the climate the linear centre first of all will be moving northward from the Mediterranean Sea north to the latitude of *London* and *Berlin* by the end of the century. At the same time, *the Arctic will gain in topological importance because it is becoming approachable for all kinds of transportation* and *because it is so rich in many of the most needed resources.* Generally speaking, the North will be gaining in topological importance, whereas areas south of the Mediterranean latitudes will be losing importance, the more so the further south they are.

Of the four problems now listed, a) *Increased heat in the warmer areas of the world* is by far most serious. Therefore the question: What can possibly help these areas? First of all, the development

> **Areas lose because of**
>
> a) *Increased heat in warm areas*
>
> b) *Oil producing areas run out of oil*
>
> c) *Decline in shipping because of Arctic routes*
>
> d) *Decline of today's linear centre*

The Linear Centre of the globe will in 100 years have moved north to the latitude of London.

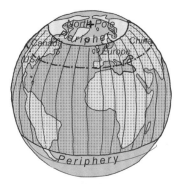

The world will experience a gradual shift from today's ribbon spatial system to a semi-sphere world with the North Pole as its symbolic centre.

of better technology, for instance for cooling buildings and workplaces. This solution to the problem depends on *holding down the price of energy,* at least in the poorer countries. The need for water is greatest in certain types of industries, including farming and the processing of cotton. Countries in the warm and arid areas of the future can try to convert their economies away from these industries to save water. *Purified water* from the oceans is already technologically possible and if a country is not too far from an ocean, there is an unlimited amount of water that can be desalinated. The problem is that this is very costly. For the very rich countries on the Persian Gulf this is today not a problem, but as the oil reserves are consumed, say in four to five decades, the lack of water could become a huge problem because of the lack of money for desalinization.

Let us now come back to d) *The reduced importance of today's linear centre and ribbon habitation around the globe.* If the warming continues through all of the 21st century and into the 22nd century, the spatial systems of the world will have changed from the spatial system of the ribbon to the spatial system of the semi-sphere with the North Pole as its centre. This *new point centre of the globe in the future* – with a floating *World Capital* – will be

very important for symbolic reasons. But there is more to this shift from the ribbon to the semi-global spatial system. The reason is that many activities will continue to be *dependent on physical nearness* to central areas and the main corridors of transportation, especially those of heavy goods and resources. *Cyberspace activities* – the Internet and global communication, on the other hand, are substantially freed from such a physical spatial system. Therefore, it would be a wise policy for nations that will be out of the physical mainstreams of the future – mostly in the far south – to concentrate on industries and activities in the area of global communication. It would also be a wise policy for these countries to create societies that are mostly self-sufficient because self-sustainable economies that are not heavily dependent on physical closeness to markets, and economies that do not need to export and import a great deal of heavy goods, can survive being out on the periphery.

In reviewing what areas will lose most in the world because of increased heat, running out of oil, their distance from the new Arctic shipping routes, or because of their worsened position in the coming change in the spatial system, the reader needs to keep in mind that the ways that are

An estimation of which areas will be experiencing low per capita availability of fresh water in 2025. Dark red means catastophically low and some most populated areas will be in a serious situation. The high North and far South have abundance of water, amo. because of how few people live there.

open to nations and areas to solve the problems lie primarily in the field of better technology and increased wealth. If nations are able to improve their standing in these two areas, many of the mentioned problems will not hit them as badly. As we have seen in *catastrophes* around the world of late, it is especially the rich that can afford the means to live comfortably, even in areas that are in many ways disadvantaged, whereas the poorer section of the population has a very hard time reducing any negative impacts, even those that are not very serious. This disparity between the rich and the poor was clearly visible when hurricane *Katrina* hit New Orleans in 2005, when the rich had the means to get away in time and to relocate in comfort elsewhere until the crisis had subsided, or – as some of them did, permanently.

Let us now *review briefly fifteen main areas of the globe* in terms of how they will be affected by the four future developments that now have been described. First the *Arctic:* The Arctic will gain in terms of all the four features – provided that the inhabitants will accept and welcome industrialization and new times. To all people that see the future in this way the Arctic areas will gain by the warming – but for those that don't like the new ways the Arctic will lose.

Let us now for clarity, recapitulate the main facts: Some of the biggest future oil and gas areas of the globe are in the Arctic, the main shipping lanes between continents will go through the Arctic, so its isolation will end and instead it will become one of the most central areas on the globe. And then finally, the shift, in the 22nd century, from today's ribbon to a spatial system that places the Arctic in the middle, will make the Arctic very important.

In terms of warming, this will be an improvement for most of *Western Europe* – most of all, however, for the northern regions in Scandinavia. The North Sea will within a few decades run out of oil, and also the south-west coast of Norway. This, however, will not be that serious for Western Europe because some of the largest oilfields of the future are in the Arctic. From there it is only a very short shipping distance into Western Europe, and in the case of gas the distances are short enough to transport it through pipelines from Western Siberia into Eastern and Central Europe.

The third area on our map is the *Mediterranean space.* This area will experience losses because of relatively less global shipping. The areas that will be most negatively affected by this are the areas around the Suez Canal. Railway lines com-

Poor people – within even a rich city – are harder hit as others as natural catastrophies occure. The photo is from New Orleans as Katrina hit in 2005.

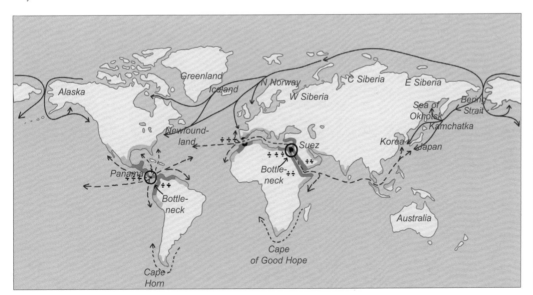

This picture shows in blue what areas will experience a relative decline in shipping once the new Arctic routes will have become very active, circa at the end of the 21st century. These are mostly areas close to the Panama and Suez Canals and the southern tips of Africa and S America.

A part of the large depot harbour of Rotterdam. Such harbours are hubs for wider distribution.

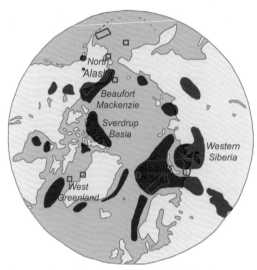

Prospective oil and gas areas in the North. Green boxes show areas of exploration drilling. The future oil areas make the Arctic important.

ing from Northern Europe, and even gas lines, will be able to substitute transportation of goods that today go by ship.

The fourth area is the *Persian Gulf*. It still has today by far the largest oil reserves, but it is also the area where oil production is by far the most extensive. Therefore many experts think that this

area will drop in oil production after only 20 to 30 years. In 40 or 50 years some areas in the Gulf may be totally running out of oil. The Gulf Area will also be negatively affected by excessive warming and dryness, and it will also, gradually, be removed from the mainstream of ocean transportation. Huge wealth is building up in this area because of oil, so those who own that wealth will be able to live in comfort thanks to refined technologies made possible by their wealth. The poor section of the populations, however, will probably suffer heavily, perhaps leading to social unrest.

The fifth area is *Russia, Siberia,* and *central Asia*. Central Asia will experience an extension of deserts but the more northern areas will be improved by the warming. Russia is the largest country in the world with 17 million square kilometres, approximately 48 times the size of Germany. The northern territories of Russia, including Siberia, will become reachable by both the ocean, rivers and new railway lines, so in the future – with so many resources rising steadily in price – Russia will become a very wealthy country. Ocean transport along Siberia is steadily improving and the growth in Russia's North has already started in Western Siberia, and gradually as the temperatures rise the development will

A scene from an oil refinery. It is strange to visualize such plants in the high North.

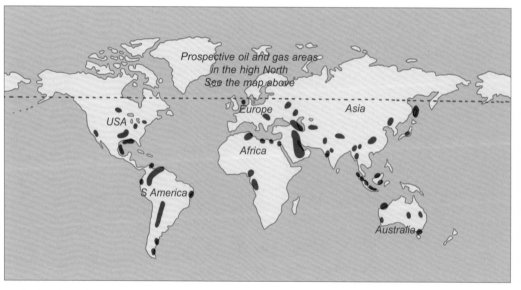

Today's oil production areas. Some of them will run out of oil in one or two decades. The area that by far has the most oil is in the Persian Gulf. Areas there, however, may start to run out of oil in about 40–50 years. New oil wells may still be found in, under or close to the areas on the map.

spread further east along the Siberian coast.

The sixth area is the *Asian subcontinent,* Pakistan, India, Bangladesh and other countries. This area will mostly suffer from the warming. There also will be, relatively, a reduction in access to global shipping. On the other hand, the area's proximity to the gas and oil resources of the Persian Gulf for 40 to 50 more years will continue to be a boon. But as it comes to transporting Central Asian and Arctic gas via pipelines, the Himalayas are a huge barrier.

Area seven is the *mainland of Southeast Asia:* China, the Korean states and other southern countries. The southernmost areas will suffer from the warming, though mostly the interior, if not at a high altitude. The reduction of oil in the Persian Gulf, Indonesia and Africa will also be a negative factor for this area in the foreseeable future, but its northern part, i.e. Northern China, will be able to substitute that partly by having gas pipelines laid from Russia. Northern China will also have access to much improved overland connections to the west with additions to the Trans-Siberian railway system.

Area eight is the *Southeast Asian islands.* These islands will not suffer heavily from the warming because of their oceanic climate and the length of the coastlines, but the reduction of oil in this area, for instance in Indonesia, is a negative factor. They will also lose relatively in terms of global shipping because of the new Arctic sea routes.

Area nine is *Australia* and *New Zealand.* In Australia, the deserts will expand but southern Australia and New Zealand are sufficiently far away from the warmest central areas of the globe not to become intolerably hot. The increased importance of Arctic shipping will mean that these areas will be at a greater disadvantage than most other areas of the globe, and also because of the movement of the linear centre of the world to the north.

Now we shift over to *the Americas,* and start in the North in area ten, Canada. Canada is the country in the world that, together with Russia, will gain most from the four described major future developments. Canada is a very northern country, except for a narrow area by the Great Lakes and on the US border. Therefore the warming will open enormous resources and provide Canada with a multitude of new transportation possibilities as the sea ice retreats from the Hudson Bay and from the numerous straits between its northern islands. In the high North there are, in the estimation of geologists, huge oil-

Extending of deserts in a warming climate is hard to work against. Overgrazing, however, is a contributing factor that needs to be curbed.

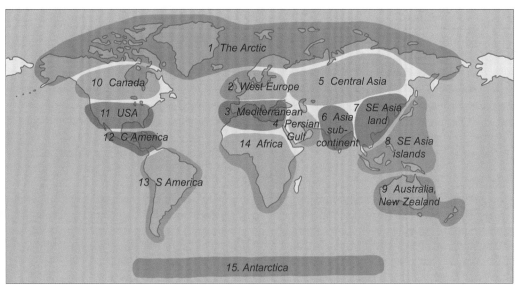

1 The Arctic
10 Canada
2 West Europe
5 Central Asia
11 USA
3 Mediterranean
4 Persian Gulf
6 Asia sub-continent
7 SE Asia land
12 C America
14 Africa
8 SE Asia islands
13 S America
9 Australia, New Zealand
15. Antarctica

The fifteen regions used in this section to signify what areas will experience losses because of: a) too much heat; b) traditional oil producing areas run out of oil; c) decline in shipping because of Arctic routes; and d) decline in spatial position because of less importance of today's linear centre.

Toronto, Canada, will increase in importance because of a warmer climate.

An open coal mine, like this one in Mongolia, is not a beautiful sight. High oil prices are now speeding up the opening of new coal mines.

Lack of electricty means that the need for wood fuel, even in desert lands, continues.

fields that will provide Canada with great wealth in the future. The change in the global spatial systems, with the movement of the main activities of Earth north, will benefit Canada greatly.

Area eleven, the lower *48 states of USA,* is on the other hand a rather southern country. New York for instance, is at the latitude of southern France. Many areas in the southwest will experience extension of deserts and extreme lack of water. Other southern areas will also become very hot. The USA is today still very rich in coal, gas, and oil, but in perhaps 30 to 40 years the oil and gas reserves will have been substantially reduced, making the country even more dependent than today on foreign oil exports. This bleak prospect already means that the USA – like India and China – *has stepped up the use of coal* for energy production, a negative decision in terms of air quality and the production of still more greenhouse gasses.

Area twelve is *Central America.* This area too will be subject to the spreading of deserts, especially in central Mexico. Many countries in this area are today rich in oil resources, but these resources are declining. An additional negative factor is that shipping through the Panama Canal will decline relatively because of Arctic shipping, even though it will be expanded to serve ships with upto 12,000 container units.

Area thirteen is *South America.* There also the deserts will be spreading, mostly along the Andes mountain range along the western ocean border. Oil is to be found there in many places today, as in Venezuela, Brazil and Chile, but oil production will be reduced. Most of all, however, the spatial situation of South America will worsen because of the northward movement of the centre of activities on Earth, which will leave the southern areas on the periphery.

Area fourteen, *Africa,* is a huge continent. The expansion of the Sahara will have very serious consequences. The areas that will be engulfed by its advancing are mainly to the south. In addition a large area in southwest Africa, the Kalahari desert, will experience an increase in desert land. Oil is to be found in many areas in Africa, for instance in Nigeria, but not in very huge quantities. As elsewhere, these areas will be badly hit in

the future when the oil reserves have been exhausted. The exhaustion of non-renewable resources usually occurs rather suddenly, meaning that the nations in question usually have not prepared well enough in advance for the necessary changes.

The living conditions at the southern end of Africa – as at the southern end of South America – will be somewhat improved because of global warming. But a negative factor for both these areas is how far south they are. This means that they will lose because some of the global shipping that now goes south of their southernmost points will be lost to Arctic routes.

Then we finally come to area fifteen, *Antarctica.* Antarctica is in some respects comparable to Greenland because both have large continental ice sheets that reach up to high altitudes. This means that these ice sheets will not melt fast. The ice sheet around Antarctica, however, will continue to break off large bergs, which will melt in many of the bays. In the latter part of the 21st century many coastal areas of Antarctica will be completely ice free. This will mean that exploitation of resources will start to be possible in just a few decades, given the unlikely result that present agreements on protecting the area as a conservation reserve are abrogated or renegotiated. Since almost no people now live in the Antarctic, and as it will remain rather cold – even with a great deal of global warming – probably not much will happen in Antarctica in the 21st century.

Most of the landmass on Earth is in the northern hemisphere and as the linear centre of activity in the northern hemisphere moves still further north because of global warming and the resulting activation of the Arctic area, Antarctica will be, spatially, the most remote point of this new spatial system of the globe in the 22nd century. What is therefore likely to happen is that people will continue to look to Antarctica as a conservation area and a sanctuary for various polar species. Nevertheless if in the late 22nd and 23rd centuries the warming still continues, Antarctica will finally be an ice-free area; a huge continent with a comfortable climate, endowed with a lot of resources, an area that some may want to migrate to in the very distant future.

3 Impact on the Global Economy

Predicting what will happen in terms of *world political development* — on which the economy of individual countries and the globe in its totality depends to a high degree — is harder than most other kinds of future predictions. Who would have thought in 1950 that the East Bloc and the Soviet Union would collapse within 30 years, and who would, in 1970, have believed that China, a country with, as it seemed to many, a completely hopeless political system and in so many ways backward — would in 30 years have become one of the world's largest industrial powers?

Global warming — and its many consequences — will without doubt have an impact on the *political stability of countries and regions* in the future, and will therefore also have a huge impact on political developments — and thus also the *global economy*. The overall picture, very generally stated, is that northern economies will in many ways gain from global warming, whereas many southern countries will be badly affected. This will undoubtedly lead to imbalances, which may, in time, be righted, but all imbalances can be very negative and can trigger wars and crises, in the world as a whole or in individual regions.

In this section on *the impact of global warming on the global economy of the future*, we are not trying, as in earlier chapters, to draw a somewhat detailed

Thinking about the future, people tend to see it in unrealistic images, as in this futurist picture of an underwater city.

picture of how some elements, for instance technological infrastructures, will develop but rather trying to translate what the various changes that loom on the horizon will mean for the global economy and the economic standing of various regions of the world.

Most people are justifiably sceptical of the *value of predictions*, even for only a few years into the future. We have already reviewed how difficult such predictions are in terms of political developments. Within this field we need to distinguish between the areas whose development is easy or hard to predict. Among the public the feeling lingers that the world is still going to experience major new inventions that will, among other things, transform the infrastructure of transportation. Some *futurists* still draw pictures of trains high above the ground and of cars that take off like airplanes, but looking back into the past, we recognize, to our amazement, that *today's traffic systems* were basically formed more than 100 years ago. Already at the start of industrialization the motorization of shipping, the invention of railways, and the invention of the automobile were all there. Of course, these means of transportation have developed and are operating more securely and faster, but basically these systems have not changed. Therefore, a prediction of how they will develop is mostly how they will be extended to reach larger areas, as has been described in this book. *Global communication* and *international information nets* are still older. The big revolution was the invention of *printing*, a field that is basically the same as 550 years ago. Already early in the 19th century, global electronic communication started with the *telegraph* in 1832 and a little later the development of the *telephone, the radio* and its extension, *television*.

The computer and *the Internet* are only one more step in the extension of these old systems. These extensions, however, are still today causing several important consequences, mostly in terms of enabling borderline people and *borderline nations* to

Prediction about political and economic development is very uncertain. Who would have believed in 1970 that China in 30 years would become a world power?

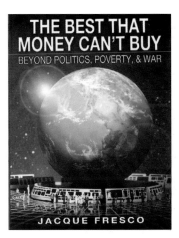

Futurist J. Fresco emphasizes that absence of war can secure the prosperity of all peoples.

Internet, computers and global communication are able to empower all individuals and nations. They, therefore, will have a great equalizing impact.

McLuhan was a prioneer in describing how advances in media are an extension of man's abilities.

enter the mainstream. Because of this, *elite peoples* and *nations* have lost some of their former privileged status. Many people have assumed that this "competition" would mean a very worsened situation for the developed world, but to most peoples' amazement the development of the third world has not — and does not need to — mean that only some nations are going to gain from these technological advances. With the use of modern communication technology all nations are developing without detracting from the meaning or power of other nations.

The high tech world requires, however, a somewhat different capacity of peoples and nations than the simpler physical world of the past. The physically strong Northerners, who were very good seamen and hard labourers, will lose some of that advantage to peoples — many of them in Southeast Asia — that are already used to living in *highly interconnected communities*. The Chinese and the citizens of India, for instance, seem to fit the new high tech types of industry very well, and they are today making big strides towards joining the *global cyberspace society of the future*.

Many economists have said that it is essential to make the jump from the *industrial economy* to the *information economy of the future*. It is certainly true that the nations that were the first to come to grips with this info-tech development have had a big advantage. Nevertheless, the whole world is catching up and every nation and almost every person can develop considerable ability in terms of computers and information technology. The belief in huge gains to be made with the new *info industries* has, however, of late started to decline because there are so many that can enter this market without much effort and at little cost. Logically, of course, the more traditional industries such as the processing of metals will, after all, remain of high importance for national economies, especially if nations can tap into earlier knowledge and skills in these fields.

In addition, what economists have of late, have been advising countries; to step back from *resource-based economies*, might be somewhat off the mark, even though such low skill, primary industries have, in the late 20th century, mostly been absorbed by low-salaried third world countries.

Increased scarcity of some basic vital resources has meant of late that resource-based economies have been getting a boost from higher prices, as is the case with the oil- and gas-rich countries of the world.

Agriculture is another industry that economists have said is *not very promising* for the future of developed nations. However today we are seeing ever higher prices for many types of food, especially if they are organically grown. In addition, agriculture in many of the traditional areas of the world — which are mostly now in warm southern areas — will suffer from global warming, not only from increased heat, but even more from increased lack of water. This will drive the price of agricultural products up.

Initially all *available water* in a country flows in its rivers and underground aquifers. This is today changing because so much water is being channeled into urban and agricultural areas. The increased demand of today's industries, urban areas and agricultural areas, means that already the *water capacity* in some areas — as in the dry northern China — is not commensurate with demand. This means that there is little or no leeway for further demand on water consumption in many industrializing areas, even without adding increased evaporation at future higher temperatures to the calculation.

In many southern countries, large rivers have already dried up even before reaching the ocean, like the Yellow River seasonally in China. This is not only serious in terms of water supply but also because it limits the other two main functions of rivers, namely, to generate electricity in hydropower plants and to provide convenient internal transportation routes. This will mean that water demanding agricultural production in the main agricultural areas of the world is bound to decrease in the future. It might even be a wise *national policy* in countries like China and India to try to *lessen their economic dependency on agriculture* as a main industry because it requires so much water, not least because the water is needed for the increasing urban areas along with modern industries as well as the need to maintain transportation routes on rivers in these countries.

Because of global warming many northern

areas that are now cold *will become good agricultural areas*. These areas are in regions like the southern part of Siberia and in Northern Canada. Temperatures may not become quite as high there as desired for most types of agriculture, but many of these areas have the great advantage of having more precipitation and less evaporation than today's agricultural areas in the South. Central Russia, like the central area of the USA, has already experienced great problems in terms of lack of water. The most famous example is that the Russians have diverted so much water for irrigation from rivers that flowed into the *Aral Sea* that the sea has almost disappeared.

The bottom line: there are new, northern areas that will be much better for agricultural production in the future than many of the traditional agricultural areas of the South. Therefore *an extensive relocation of agriculture* on the surface of the globe is to be expected in the future. The scarcity of agricultural land in the developed countries – where most urban areas have been built on the best agricultural lands – will contribute to this restructuring.

Now, some of the readers of this book would like *to remind the author to be more pessimistic* about the projected weather changes, underlining that the changes are going to mean much difficulties for many traditional industries like farming and tourism in some countries, and that some of these countries will, in the next decade, be hard hit and suffer the results. In response the author wants to remind the reader of *the impact of the passing of time*, as was described on page 37. If we look at the world as something *static* and something that is *not meant to change*, then the picture is bleak. If, on the other hand, we accept that *everything is constantly in a process of change* and that no nation or region will or can remain static and unchanged for many decades – with or without the impacts of global warming – we can get *into the necessary frame of mind* to be *willing to embrace the changes that are coming upon us*.

It is certainly true that *periods of change in the world's history* can be very trying, but it is also true that these trying times have often produced unexpected *forces of innovation* and *a will to survive and persist*. In only about 10 years after World War II

Germany was already back on its feet economically. It has even been said that the eradication of old schemes and old factories in Germany was the key for the rapid introduction of modern ways. The same can happen in countries that will be hard hit by some impacts of today's or future global weather changes. If they have the will, the power and the imagination needed *they can turn these catastrophes to their advantage* and maybe even surface – as Germany did after World War II – in a surprisingly short period, even become more powerful than before. Looking at the problems of global warming in this way, we have the possibility *to interpret the upcoming changes* not as *disasters* but rather as *opportunities* to redefine and reform our ways; an opportunity for *creating a better world* and *stronger economies*.

Most problems of today are caused by rather *unintelligent ideological conflicts*, for instance between the Muslim and the Western Worlds. One could even say that this is a "luxury" that only a very rich world can allow itself, i.e. to spend all this money on war machines and warfare that are no good for anybody. It is now pouring over us that the world, or certain parts of it, will be hit in a colossal way by the impacts of global warming. This may possibly lead the world population as a whole to understand that *we need to cultivate what is common to humankind* rather than *what separates us into factions* that produce problems that result in open conflicts.

The migration of the future from problem areas in the South to the North – which needs to happen to some degree – can only happen if we stop the hatred that now has developed between, for instance, the *Southern and Central Muslim World* and the *North-western Christian World*. This hatred is probably in fact not so much hatred of Western nations as an *intense dislike of many modern ways* that developed first in the West, as for instance the emancipation of women. These modern ways are developing in almost all countries, countries as unlike as Japan, Kenya and Greenland – even though grudgingly received among by the elders in most countries. The integration of all cultures into *one common global culture* is bound to happen, and many will regret the old times and ways, as for example the author of this book. What the

"It is now pouring over us that the world, or certain parts of it, will be hit in a colossal way by the impacts of global warming. This may possibly lead the world population as a whole to understand that we need to cultivate what is common to humankind rather than what separates us ..."

McDonald's, although originaly American, is only one appearance of an international culture.

The retreat of snow and ice in the Arctic makes it possible to access its huge minerals, oil and gas riches.

author, however, sees as positive about this *integration* is that such an integration seems to be totally essential if the *mega-scale migration processes* that this book talks about are going to happen in a somewhat peaceful manner. If peace and acceptance about certain *common ground* rules can not be achieved, there will be a huge reluctance to open up the southern borders, for instance in Europe and Russia, to allow the necessary northward migration.

Now many people may say: Why should the European Union and Russia open their gateways into the southern areas? The answer is: "It is a moral duty". We cannot prevent people from these *disaster areas of global warming* from moving to the more benign environments and climates of the North. A still greater reason for this moral necessity is that most Northern nations are not in lack of living space because as the globe gets warmer, huge northern areas of Scandinavia, Siberia, Northern Canada and Alaska will become very fit as areas to migrate to. To put it briefly; people should do as the bio-systems, the birds and the plants in these southern regions do: *move north.*

In the history of the world – be it because of cooling or warming – ecosystems and migrations have always responded to climate changes. Humans have always been a part of a biosphere and they have followed their prey, fish, animals and plants, to the north in times of warming. One ordered way of letting this happen is to extend the European Union to the south. In view of the terrible conflicts that are waged today between the Western and Muslim worlds, this might sound absolutely unthinkable. Still, one only needs to look a short time back in history to see how the main enemies 60 years ago in World War II – Germany, Japan and the USA – have become best friends. Similarly, the main enemies of the Cold War in the 1950's and 60's – Russia and the USA – have become friends. We can also think about 25 years back, in terms of the Persian Gulf region, when the USA and Iran – with the Shah at the helm – were the best of friends. Obviously, if we only think in terms of the very short scale of months and years, we see no possibilities. If, on the other hand, we broaden our

horizon and study even recent history we see that conflicts that now seem to be unsolvable, with the passing of time are resolved.

Judging from the lesson *that disasters,* or looming disasters, *teach people to emphasise important truths and relations* and to shed trivialities, it is possible that the upcoming disaster of global warming may provide the tools for making people understand that they *cannot isolate themselves from the unifying tendencies of the emerging global culture.* If the various peoples and cultures of the current world do not adjust, they may end up pushed to one side in their own hot and arid areas, isolated in their hostility to the rest of the world. So global warming therefore actually could mean a necessary catastrophe to push the peoples of the world towards forming *one common global culture* and *one common humane world.* This process could greatly reduce the numbers of armed conflicts and the need to finance the production of expensive armaments and maintain sizable armies.

The catastrophes of global warming that the world is starting to experince, in fact need exactly the money that now goes to support man made warfare to combat instead the catastrophes, the "nature warfares" that are coming upon us. The nations of the world need to unite in helping the regions that will be hardest hit, for example by allowing *environmental refugees* to migrate to their countries. Hopefully the warming will stop and thus also the troubles of these trying environmentally-based adjustments. But if the warming continues we need to transform today's discordant global culture to a *single integrated world.* If we succeed in doing that we will be able to solve most of the problems that are coming up on us with the changes in the climate of the world.

Some of *the largest tasks of our world* in maybe the next 100 years, will also include physical tasks, such as a major relocation of agricultural areas to the North. But most trying, culturally and politically, will be the *relocation of large numbers of people* from areas that become uninhabitable because of various global warming related reasons, to areas that are going to become very comfortably habitable – mostly in the North.

The Aral Sea in Russia has dried up because of too extensive irrigation projects.

4 Impact on Geopolitics

We have now assessed, in the last three sections, the impacts of *the picture of the world in the future* that has been drafted in this book. We started by assessing what areas would gain and what areas would lose because of global warming. Then we assessed the impacts of the opening up of new arctic oilfields, and new northern transportation routes, as well as the impacts of the semi-spherical spatial system of the globe in the future.

In the nearest future of 10-50 years, the development of the oilfields in the Barents Sea – mostly for the USA and West European markets – will be the most critical aspect in terms of *the structure of geopolitics* in the 21st century.

The conflicts between the West and the Muslim world – where most of the traditional oilfields are located – is leading to speeding up the opening of the new oilfields of the Arctic, and it is the retreat of sea ice, because of global warming, that is allowing this to happen. The benefits of these developments are many. First of all, the predicted shortage of the supply of oil and gas is further in the future than has been assumed of late. Secondly, the opening of new major oilfields in the Barents Sea will also mean that the increase in world oil production will mean that pressure for higher prices will probably decrease in the next few decades.

As the West will not be as dependent on Arab oil because of the new Arctic oil fields, the conflicts and tensions in the *Persian Gulf* region and the near East will ease considerably. The reason why the West has been engaged so heavily in the affairs of the Gulf region is the necessity of assuring access to the oil fields in the Gulf. This has for instance meant supporting or installing governments positive to the West. To maintain armed forces in the region reduces the danger that groups that are negative towards Western views and influences will block oil exports.

One of the reasons for the rising oil prices of late has been the *uncertain geopolitical situation* in the Persian Gulf region. Therefore, hopefully as early as 2010 or 2015, as the West has started to become less dependent on oil from the Persian Gulf, the geopolitical tensions caused by the threat of oil shortages will lead to the easing of the West-Muslim conflicts, and oil prices come down. Therefore, as shown on the map at the bottom of the column – where the West is supplied with oil from the Barents Sea instead of having to transport a large share of it for long distances through problematic areas – the geographic changes in source of supply and delivery routes will be factors that will *change the geopolitics of the world* in a fundamental way.

As seen on the map to left, *new political partnerships* will then develop because of new interrelations of nations in terms of oil. We have already explained the emerging North Atlantic-Russian partnership. The area at the bottom of the map shows the closer linking of the Persian Gulf to the Asian subcontinent and Southeast Asia as one partnership in the future, as the West gets more out of the picture.

Thirdly, a partnership between Russia and China has started to develop because Russia will not only be able to supply the West with oil and gas but China as well, most importantly with gas, through pipelines running through Central Asia.

The *retreat of the Arctic sea ice* is not only open-

The West–Muslim conflict – mostly because of oil interests – is shaping today's geopolitics.

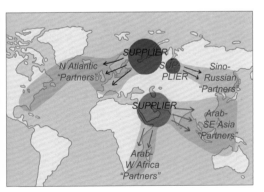

Once the West has been begun to be mostly supplied with Russian oil, the division into new geopolitical oil partnerships becomes clearer.

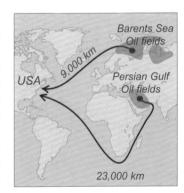

The dependence of the West on Persian Gulf oil will be resolved by the opening of Arctic oilfields.

119

ing up new oilfields but also various other types of resources along the Arctic rim and also in *the huge flat continental shelf* that is currently under the Arctic Ocean. This benefit is added to the fact that the shipping distances between the Arctic and the main markets in Southeast Asia, Europe and North America will become much shorter. For the northern nations of the world, it is of high importance not to have to rely solely on the shipping routes through the narrow canals of Panama and Suez in the South, where episodes of political instability are likely to erupt in the future. The opening of the Arctic sea routes – where the ships can go "north of conflict", is therefore very valuable geopolitically, especially for the North.

This *sailing north of conflict* actually happened *in World War II,* as the USA became the main supplier of Britain in the early years of the war. In this case the sea routes from America went north of the conflict areas of the Atlantic, i.e. up to Greenland and Iceland and from there to Britain. Later in the war, as the Russians started to mobilize to fight the Germans, this route over Iceland – north of the areas in the Atlantic that were largely controlled by the Germans was totally essential for the Allied victory over Germany.

Deep fjords in Iceland were perfect to keep Allied warships secure from German air raids in WW II. In Norwegian fjords German ships had shelter.

This is because it was mostly through US equipment that the Soviet's Red Army was able to fight the decisive battles of the war along the Eastern Front, with the Soviets losing most people during the war or about 20 million people.

In the ensuing *Cold War* the picture of friends and foes was totally reversed: the earlier friend Russia now became the main enemy and threat to the USA. In the Cold War the ocean area between Iceland and Murmansk again became an area of conflict, now between the USA and the Soviet Union, as shown on the map on page 74.

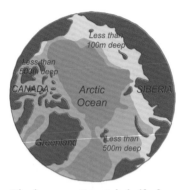

The huge continental shelf of the Arctic Ocean offers extensive possibilities for oil extraction. Waters of 100m or less (light blue) are extensive along the Siberian coast. Elsewhere areas 100–500m in depth (middle blue) are also extensive.

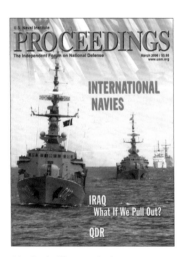

Navies will grow in importance in the Arctic because as the ice retreats more coasts get exposed.

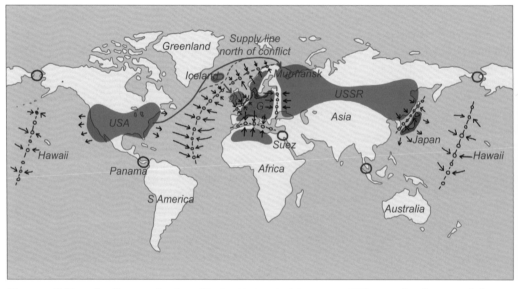

The possibillity of sailing north of conflict in the North Atlantic in WW II proved to be one of the main reasons why the USA could supply Britain and Russia with necessary equipment to defeat the Germans. Allied forces in Iceland were key to be able to control the North Atlantic sufficiently well.

Now, let's start to draft a picture of what the future is likely to may hold in terms of geopolitics. The two maps on his page show the areas the author thinks will be *the main conflict areas of the world in the 21st century*. These areas are the Persian Gulf, the area between Russia and the Muslim world, the border between Russia and China, where China is probably going to claim areas for its needed northern expansion, and areas north of Japan: the Sakhalin and the Kuril Islands that Russia acquired as the victor's booty after the Second World War. These areas and the Arctic are becoming increasingly important because of global warming and because of oilfields.

The map also shows the whole Southeast Asian area as a possible conflict area, simply because of the dynamism and the uneven economic developments in the area. *A disturbance in a balance of power very often leads to political conflicts.* As we look at these future conflict areas in the South, we realize better what luck it is that the Arctic Sea routes are opening up. The possibility *to sail north of the conflict areas* will be of benefit for the whole world because a more secure global transportation in times of crisis is very important. If world shipping is disturbed the globe will collapse into a worldwide economic crisis.

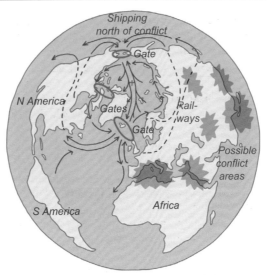

This map shows better than the map below in what way the Arctic routes will allow global shipping to take place north of conflict.

One of the main tools available to reduce conflicts is to *integrate regions* and *abolish borders*. In this way areas *become interdependent* – i.e. they become one area in terms of the economy, as well as in other ways. This was the main thinking behind the establishing of the *European Union* in 1957, and because Western Europe is now an area of com-

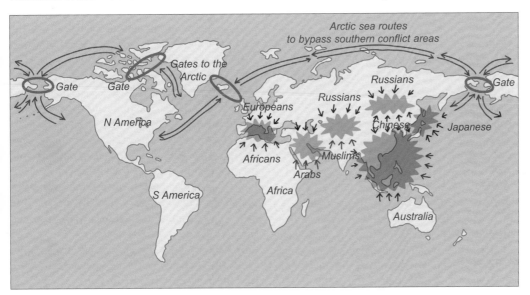

The author's prediction of the main conflict areas of the world in the 21st century. These areas will mostly be in the Asian sub-continent and in South-east Asia. The lifelines of global shipping go through there now. Arctic routes will allow global shipping to go north of these areas.

We tend to forget that old wrong-doings and hostilities linger as a potential source of conflict.

If the Arctic rim becomes warm, and filled with activity, a strong union of the bordering nations may develop. The picture shows the Russian Arctic harbour Murmansk that played a central role in World War II.

Russians own by far most ice-breakers. They are key to being able to operate in the Arctic.

mon interests there is no more reason for, say, the French and the Germans, to go to war over border areas, conflicts that were the main cause for the wars between these nations in the two World Wars of the 20th century.

This scheme of *creating large interdependent regions* needs to be applied in all areas of the world. This scheme of interdependence is the very best tool the world has at its disposal to become a peaceful world.

One of the main reasons for the insecurities of nations, and a possible source of conflict, is if the borders are not natural borders and secondly, if the borders are very easily penetrated. The main reason why the USA has not experienced a foreign war on its own soil since World War II is that it is separated from most of the rest of the world by very effective barriers: the Atlantic and Pacific Oceans and the Caribbean Sea. Relations with Canada to the north are peaceful and the cold and ice of the Arctic are protective barriers for both the USA and Canada. Modern technological capability and terrorism, however, is currently seen as a threat to the USA's earlier relative isolation, as demonstrated by the destruction of the World Trade Towers on 9/11 2001.

It holds generally true for all the eight countries bordering the Arctic Ocean that *the northern borders* have, up to now, *been totally secure* against almost any intrusion or threat from terrorist attacks, illegal imports and unwanted immigration. This is very different from the southern borders of, say, Europe and the USA. This picture of the Arctic might change fundamentally as the sea ice starts to disappear from the Arctic Ocean and these long coastlines start to open up for every seafarer.

The increased military and coast guard expenditures will *hit the Russians by far the most*. They already have a long hostile southern border and a potentially vulnerable eastern sea border. The border with Western Europe, on the other hand, one hopes, is becoming rather calm, but this new addition of a very long accessible Arctic coastline will increase Russia's vulnerability.

The coastlines of Northern Canada are also extensive, especially because of the very many islands, whereas the northern coastlines of

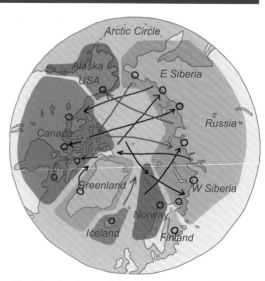

The circular form of the Arctic, the area's huge resources and common interests e.g. in terms of security, will bring the Arctic countries together.

Greenland, Alaska, Norway, and Sweden are much shorter. Their opening, however, is going to present new concerns and responsibilities for these nations in the future.

What is of special significance in this perspective is that the Arctic Ocean is *a circular body of water* that is *rather closed off* from the rest of the world. First of all there is only a very narrow gate through the Bering Strait from the Pacific into the Arctic Ocean. The gate from the Atlantic into the enclosed space of the Arctic Ocean – between Greenland, Iceland and the UK – is much wider but there are many possibilities of building up monitoring and defense mechanisms. Such a system was installed by the USA during the Cold War. It consisted of radar stations and submarine listening devices that, however, have mostly become dysfunctional as the USA left its post in Iceland in 2006. Such a surveillance system is again becoming very important because of the increased oil shipments and the need to guard this gate to the Arctic space.

To enter the Arctic Ocean through the Canadian Archipelago is not easy because there are only narrow straits that are rather easy to control. There are only eight nations bordering the Arctic Ocean, and all culturally related. The thawing of the Cold War in the 1980's allowed them to start holding meetings on the Arctic. This led to the *establishment of the Arctic Council in 1996.* Subjects of common interest include the presence and development of oil reserves.

It is quite possible that this ocean will be defined as *an inland ocean* of the Arctic rim countries, where the entrance will be guarded in common by the Arctic nations. It is possible that NATO (the North Atlantic Treaty Organization) will be expanded to cover this task, but that would mean including Russia, Sweden and Finland in NATO.

Another possibility is that the Arctic Council will form *a military co-ordination unit* to devise a plan to keep this area free of problem makers, for instance by heavy surveillance at the narrow gates into the Arctic Ocean.

If the warming continues in the 22nd century, it is quite likely that there will be *increased interrelationships* developing among the various opposing regions around the circle of the Arctic Ocean. This will mean that the Arctic Ocean will become somewhat *similar to the Mediterranean Sea earlier.*

We have now, for a while, been discussing the positive impacts of the Arctic Ocean becoming ice free, including the new and shorter global shipping routes. Let us now turn to what will happen in the land areas *in terms of geopolitics because of global warming.* The subject that is by far the most important in connection with this is how *the northward migration of people* will take place in the future. In order to try to start to understand how this will happen it is necessary to look at a few maps that show what are the *main obstacles* or *barriers* to this *northern migration,* primarily in the northern hemisphere.

On the map below, *natural horizontal barriers* to a northward migration are shown in colour, brown for deserts and mountain ranges and blue for water. *In the Americas* we immediately recognize that the main obstacles to migration – the vertical mountain ranges running along the whole western part – are obstacles to east-west but not northward migration.

In Eurasia, on the other hand, there are *very difficult horizontal barriers* to northern migration, both in terms of mountain ranges – the Alps and the Himalayas – and bodies of water – the Mediterranean and the Baltic Sea in the northern hemisphere and the division of Australia and New Zealand from Asia by a wide body of water in the Southern hemisphere.

We have now seen that in North America there are only very insignificant natural barriers to

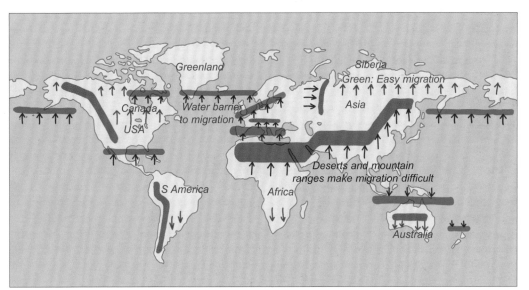

With excessive global warming huge numbers of people in hard stricken southern areas will have to migrate north (or to the far South). The natural, horizontal barriers to migration – deserts, mountain ridges and oceans – are different in the various regions of the world, as this map shows.

The number of environmental or climate refugees will grow immensely in the future.

People can go on small boats between Europe, Africa and the Near East. Here, the famous Exodus of Jews from Europe to Israel.

northward migration. In addition the political border of the USA and Canada is not of major significance in terms of a possible northern migration. In the case of the border of the USA with Mexico, in contrast, there are already illegal immigration difficulties that the USA wants to get under controle. This pressure to go north will increase with still more global warming.

As for Eurasia, the Mediterranean Sea is the biggest separation between Europe and Africa. Slightly farther north the Alps present, even today, a major barrier. This barrier will probably mean that the people who cross over the Mediterranean will mainly remain in the area south of the Alps – in fact, a continuation of the present often thwarted attempts of Africans to seek a chance for a better standard of living in European countries.

In the Near East it is the Bosporus Strait and the Black Sea that are the biggest barriers. With the Himalayas we have a barrier that is probably the most serious for the idea of solving global warming problems by migration. First of all this mountain range is very high and wide so that tunnels, as in the Alps, are hard to build. In this area there are also political and cultural differences that will make northward migration very difficult. Also certain areas north of the Himalayas are going to be deserts.

The former republics of the Soviet Union – Kazakhstan, Mongolia, and Uzbekistan – are probably going to benefit from their earlier membership in the Soviet Union in terms of migration – even though they are today independent from Russia. This old relationship will help as it comes to the establishing of an economic union that abolishes borders in this area.

These Central Asian areas are very sparsely populated but, at the same time, areas that have much potential in terms of resources and because of increased accessibility in the future.

In China the problem of its North is the difficult mountain areas. They will, however, improve in terms of habitation, despite the rough terrain and increasing lack of water. There are some problems connected to a high altitude that need to be realized when assessing the potential of high terrains. The most important is that they lack

the benefit of being close to the coast as is much of lowland China. In addition, most low areas enjoy the benefits of the accumulating water from the mountain ranges, flowing in calm, big rivers. Therefore most of the huge amounts of water essential for cities, industries and agriculture, are most available in low areas.

In many of the mountain areas, on the other hand, there is a lack of water, for instance in the Gobi Desert. In addition, mountain areas are rugged and do not have the benefit of easy transportation on calm rivers and most often lack sufficient flat terrain for building railroads and highways.

In the picture to the left below we see that huge settlements, in our modern times of *urbanization,* need not take up much space. And, as a matter of fact, *a high population density* is actually needed to create vibrant economies and transportation systems. The high-density urban areas of today have needs that are very different from earlier times when nations were spread over agricultural areas – living in more or less self-sustainable units.

We can see the benefits of dense living in city-states like *Hong Kong* and *Singapore*. There we have a minimum distance for all transportation and the authorities don't have *the burden* of having *to pay for infrastructure* stretched over a wide, uneconomical hinterland.

Today the areas that nations used to need for food production and resource exploitation *need not be in the same areas* where the people live. The absence of the need for an attached hinterland may be still further enhanced in the future – perhaps even to such a degree that the *agricultural areas* that are needed to feed, even the Indians and the Chinese, will not be solely in the now ever drier river valleys of these countries, but may, to an increasing degree, be located in the high North where there is enough water for irrigation of fields. The *ampleness of water* means that, in spite of this, there will still be enough water left in the rivers so they can continue to function as transportation corridors.

Mega cities like Hong Kong show us how well cities without attached hinderland can thrieve.

6 The North: The Future Area of the Globe

1 On how Warm the North will Get

As has already been described in previous chapters (see pages 20 and 32), the Arctic will experience more extensive warming in the future – probably by 4 to 7°C in the present century – than other areas of the globe. This will result in a rapid reduction of the polar ice and create a more habitable climate that will make *the North* the *premier future development area of the globe*. This 6th and last chapter of the book will show in some detail how the activation of the northern part of the hemisphere will happen. In chapter 3, on pages 71 and 72, five steps in *how shipping will evolve in the Arctic* were shown on maps – the map to the right being one of them. This shipping development will be a decisive factor in what steps are taken to utilize the enormous resources of the Arctic Rim and the Arctic Ocean. This development has already started in the huge oil and gas fields of Western Siberia that have recently started to expand into the Barents Sea north of Siberia and Northern Norway.

The biggest step in this development, however, will take place as this shipping route into the North Pacific Ocean will become *open all year round*. In the era of thinning ice, to start with, this

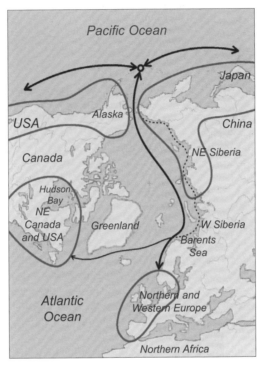

The retreat of sea ice in the Arctic will open easy shipping routes between the Atlantic and the Pacific in the late summer, within 20–30 years.

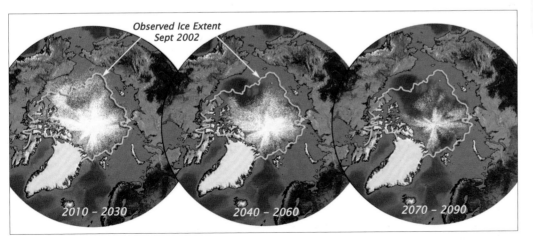

The decline of sea ice in the Arctic in this century in three steps. The pictures show the September status of the ice (the year's least ice), compared to the little sea ice of Sept 2002 (white line). Some models predict disappearance of summer sea ice by 2100. (The map does not show other changes).

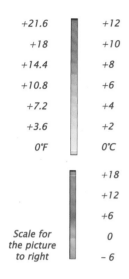

+21.6	+12
+18	+10
+14.4	+8
+10.8	+6
+7.2	+4
+3.6	+2
0°F	0°C

	+18
	+12
	+6
Scale for the picture to right	0
	−6

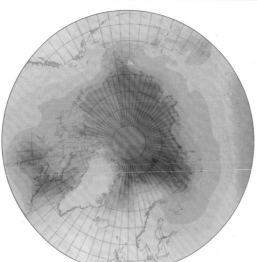

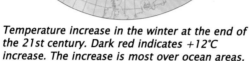

Temperature increase in the winter at the end of the 21st century. Dark red indicates +12°C increase. The increase is most over ocean areas.

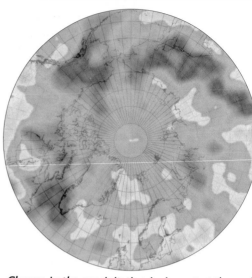

Change in the precipitation in August at the end of the century. Green indicates increase up to 18mm, but brown shows areas of decrease.

will be realized by building containerships and tankers somewhat like icebreakers. These huge strong ships would run regularly *between depot harbours* at each end of the Arctic Sea route – possibly located in Iceland and Dutch Harbour in the Aleutian Islands.

Before we embark on the description of the North as one of the most important future areas of the globe in sections 2, 3 and 4 of this chapter, let us first start by reviewing the discussion *on how warm the North will get.* This is absolutely essential because what will happen in the Arctic is totally dependent on how warm it is going to get there, not only in this century but also beyond.

Of course the accuracy of the projections of the warming in the 21st century is a subject of scientific and political debate. Most scientists today, however, agree that the *anthropogenic greenhouse gas emissions* are the main cause for the increased temperatures. The debate today centres more on *whether it is possible,* realistically, *to stop the warming.* The intent to stop the increase or reduce emissions is being worked on under the banner of the *Kyoto Protocol,* but the emissions of the nations that signed the agreement only account for about 25% of the world total. The nations that have signed are mostly developed nations that have already put *the period of heavy industry* behind them

and want to push the remains of heavy industry beyond their borders by accepting limits to their carbon greenhouse emissions. In spite of this, most of the Kyoto signatories have not been able to meet their targets.

In January 2006 another group of nations meeting in Sidney, Australia, presented a somewhat different program to deal with the emissions. This group is called *The Asia-Pacific Partnership on Clean Development and Climate,* A6 for short. The six nations of the partnership are the industrial giants – the USA, China, Japan, South Korea, India, and Australia – that today account for half of the world emissions. The main characteristic of the A6 approach is *not to adopt quotas* but rather to *rely on technological advances* and *free market mechanisms* to achieve "relative" reductions in undesirable emissions. The word relative is necessary as a qualifier because countries like China and India – in common with other Third World countries – are *only in the starting phases* of their *industrialization* and *urbanization,* two processes that are bound to substantially increase emissions of these countries, which account for about 80% of the world population.

A major issue in the discussion on global warming is the possibility of *increasing the efficiency of energy use* since up to 90% goes to waste because, among other things, of inefficient machinery and poor insulation of buildings. Important advances have been reached, for instance within the automobile industry. But in most cases the mechanisms and schemes needed to achieve substantial savings are too complicated and costly to be realistic.

Another field where important scientific work is under way is the *development of less polluting* and – if possible – *renewable ways of producing energy*, for instance from solar and wind sources. Unfortunately, there are several serious environmental impacts from these technologies including visual and noise pollution. Therefore, a sustainable and clean energy solution is not yet in sight. A part of the public discussion is theories that outline possible *changes in the conveyer belt* of the world ocean currents (see the picture below). One theory says that the belt will slow down, meaning that less warmth would be transported from the warm areas of the globe to the cold ones. A part of this belt is the warm *Gulf Stream,* which originates in the Gulf of Mexico and transports considerable warmth to the North Atlantic and into the Arctic Ocean, where it now has better access because of the retreating of the sea ice.

Another theory says that the Gulf Stream may branch out differently, meaning that some countries will get less heat and other countries more heat. A third theory is that the melting of the Polar and Greenland ice will mean that the currents coming out of the Arctic – mostly through the Denmark Strait between Greenland and Iceland – will have a lower salt content. The lower salt content would mean that these currents would not sink as fast by Greenland but float farther south on the surface of the ocean, passing along the North American Coast. This would mean cooling of that coast, as described in the science fiction film *The Day After*.

The problem with these theories of how the currents of the world oceans may change is that *real data to verify the predictions* of the theories or models *are largely missing*. Such data would need preferably to be decades or centuries long *time series* on changes in temperature, speed or the salt content.

The model of how the temperatures of the world climate will, and have been changing can be tested against extensive worldwide temperature measurements that extend up to two hundred years back in time.

The measurement of the various qualities of ocean currents are now being increased, and in a few decades we will have real data to verify or reject what the many ocean current models are now predicting. Until then we can hardly include these theories in the public discussion.

Objections

Some will say:
Let's stop global warming!
Sadly, not possible -
But it saves resources.

Others will say:
Let's create clean energy!
A good solution -
But not yet in sight.

As of now,
The warming will continue.
Let's draft a picture
Of northward migration.

Migration is a natural way
To adjust to warming and cooling.
Only static thinking
Makes it seem impossible.

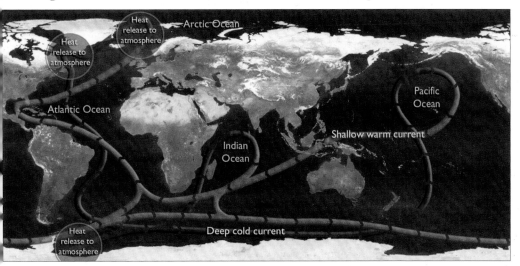

The circulation pattern of the world's oceans. The warm currents absorb heat in the warm areas and release it to the atmosphere in the cold areas. There are theories that postulate that this conveyor belt may slow down or otherwise be disturbed, but data to verify these predictions are not available.

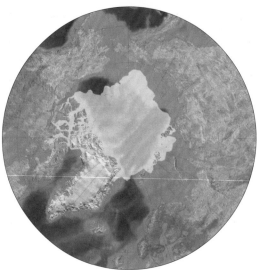

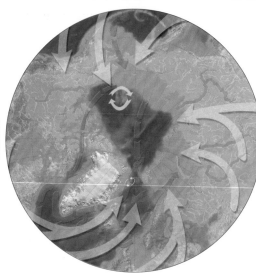

Observed sea ice cover in September 2002. Little ice in Barents Sea allows the Gulf Stream to go further north and ships to sail even to the Pacific.

Red and blue arrows: Warm and cold currents. Light blue arrows: Direction of wind flow into the Arctic, and a gyre in the Arctic Ocean.

The large *Arctic Climate Impact Assessment* (ACIA) *report* (see page 19) that describes how the climate of the Arctic will develop in this century – and what the impacts will be, mostly on the natural environment and indigenous communities – does not include any of the theories of changing ocean currents in its climatic scenarios. Therefore neither does this present book. Rather, it builds on the ACIA report that predicts that during this century most of the Arctic sea ice will retreat step

by step in the summer and that temperatures will rise by 4 to 7°C.

The projections of the ACIA report include the assumption that during this century some success will be achieved in some areas of the globe in slowing the increase of greenhouse gas emissions. This, however, will not be the case in many areas, so that the warming will continue. The assessment of the ACIA report builds on so-called *A2* and *B2 emission scenarios* provided by the

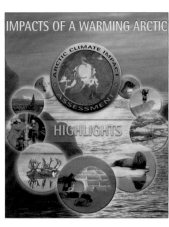

The Arctic Council published the Arctic Climate Impact Assessment (ACIA) in 2004.

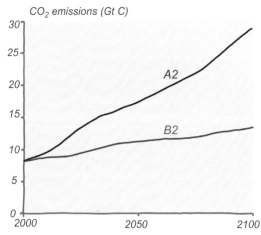

The A2 and B scenarios on CO₂ emissions to 2100. The projected temperature rise follows these CO₂ scenarios closely. (See the right page).

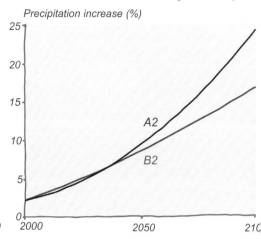

Projections of increased precipitation in the Arctic follows the projected increase in temperature very closely.

IPCC. These are in the middle range of the scenarios. According to A2, Arctic temperatures will rise by 8°C and the global mean by 3.5°C in this century. The numbers for B2 are 5°C for the Arctic and 2.5°C for the globe.

Almost all projections on increase in temperatures only go to 2100. The reason for this is that the uncertainty of the projections increases by time. It is for example possible that the world will develop ways to decrease emissions in this century, and it is also possible that it will not, so the increase will continue with the industrialization of the third world.

Looking at the graph below we see that the projection lines point to a still further increase in temperatures after 2100. The author of this book took the liberty of – unscientifically – extending the A2 and B2 projections for the Arctic to an additional 50 years; to 2150. By only adding 50 years, the range of the projection for 2000 to 2100 almost doubles; i.e. to 8 to 13°C. The picture of the future of the globe and the Arctic

depends mostly on how high the temperatures will get – and there we should not only look to this century. The author makes the basic assumption *that the temperatures will continue to rise substantially beyond 2100* – and the descriptions of the future of the world and the North in this book are based on that assumption.

Lately people have started to be willing to accept the idea that, in spite of continued efforts under the *Kyoto pact* and the further development of cleaner technologies, mankind must live with the unfortunate fact that as of now, we are not capable of stopping the warming of the climate. This means that we have to s*tart studying more seriously* how we can *adapt to the changes in climate*. This book is meant to be a contribution to help us finds ways to do so.

Many theorists developing this *adaptation strategy* only look on measures that can be applied locally. This is for instance the main thrust of the measures proposed by *Frances Cairncross,* the president of the British Society for the Advancement

"The author makes the basic assumption that the temperatures will continue to rise substantially beyond 2100 – and the descriptions of the future of the world and the North in this book, is based on that assumption".

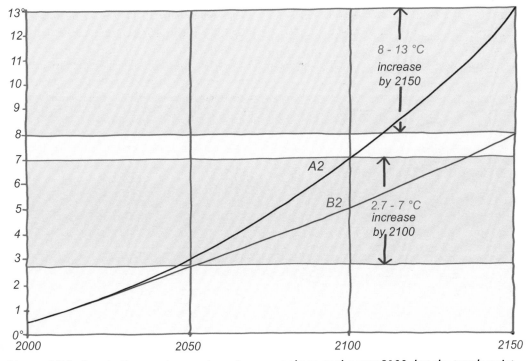

Most published projections on temperatures increase only go to the year 2100, but the trends point to a further increase after 2100. Here, 50 years are added to a scientific projection from the ACIA report. Such a continuation would lead to doubling of temperatures in the Arctic from 2100 to 2150.

Increase
in number of
growing days

+70
+60
+50
+40
+30
+20
+10
0

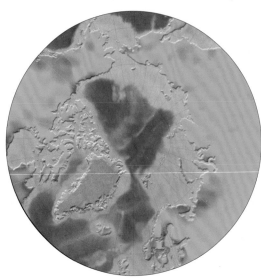

Projected increase in the number of growing days in 2070 to 2090. A two month increase (redish colours) is mostly projected at coasts.

of Science. Ms Cairncross proposes, for instance, *better flood protection* along coasts and the *establishment of migration routes* for wild animals from south to north.

This book points out that we can also, like animals, birds, fish, and plants, *migrate* to places that better fit our needs; that is, to leave places that will experience worsened climatic conditions to find new and better ones. The *static* cultural and societal *structures of today* are a very serious obstacle to such human migration in many regions of the globe. But, nevertheless, if we make a genuine effort, solutions can be found, as this book has already shown.

The *main novelty,* however, *of the present book* is its overall urge *to embrace the changes* that come with global warming in ways that allow humankind to gain from the changes.

These are just first drafts about what we may experience in the future in terms of global development. As of now, we do not know for certain how things will proceed, but the picture will start to become clearer as this century progresses.

Strangely, there is *only one simple question* that we need to have an answer to in order to make rather accurate projections: What will be *the temperatures* in the world and in the North in the future? If the temperatures remain the same, not much will happen. If they rise by 4 - 7°C, the impact will be colossal, but if the warming continues much beyond that, *the North* will indeed become *the future area of the globe,* as described in this book.

In Europe routes have been created to allow wild animals to migrate over man-made barriers.

In the future natural environments in many places of the Arctic will look very much like that of central Scandinavia today, as is shown in this picture. Scandinavian environment will, at the same time, develop to be like that of central Europe. In short: The climate and biozones spread north.

2 Exploration, Research and Organizations

At the end of each *ice age*, migration to the North started again with birds, fish, plants, and animals leading the way. The *Paleo-Eskimos* and the *Inuit*, whose culture and physiognomy are most fit for cold climates, probably most often followed before long. This is very little recorded history and there have been relatively few archaeological digs to flesh out any details of these northward migrations. In fact, each region and each culture involved has its own history about how it explored its northern regions. The *European history of Nordic exploration* is commonly said to have started with a journey by the Greek navigator *Pytheas* around 330 BC. He sailed north-west from Scotland seeking a legendary land called *Thule*, where the sun shone all night long. Many scholars think that the country that he discovered was *Iceland*.

In the 8th century AD Norsemen discovered Iceland and *Ingólfur Arnarsson* is said to have been the first settler in 874. The discovery of Iceland seems to be the first advance of Europeans that later led to the new worlds of the Americas. This

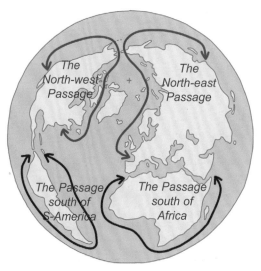

The Americas and Eurasia–Africa are two islands. The effort to find passage south and north of them has therefore been of great interest.

exploration and settlement process continued with *Eric the Red,* who lived in Iceland and established a colony in *Greenland* in 986, a colony in its south-western part that survived for several centuries. Around the year 1000 another Icelander, *Leif Eriksson,* sailed to *North America,* which he called *Vinland.* He and other Icelandic-Greenlanders visited this area several times, but the single fixed settlement at L'Anse aux Meadows did not survive for long. It was not until 1492 that the voyage of *Columbus* opened the way to the settlement of the Americas.

In the 1490's *John Cabot* proposed that there must be a direct way to the Orient via the *North-west passage.* Another theoretical possibility to reach the Orient was to go over the Arctic via the *North-east Passage,* a route that the Europeans have, for a long time, been dreaming of. As it turned out the routes to the Orient south of South America and Africa were discovered earlier and were immediately navigable.

In the middle of the 19th century the interest in the *Arctic* and *Antarctic* increased immensely, among other reasons because of better navigation instruments and better ships for exploration. In 1879 and 1880 the *first international Polar conferences* took place, promoting the idea of regional economic and scientific co-operation in research and exploration. In the late 19th century, there developed a huge fascination for such journeys of exploration into the unknown. This was partly fuelled by the interest of nations in staking their claim to areas that might be of value in the future.

Several voyages of exploration were instituted into the Arctic, most of them ending in huge tragedies as ships were crushed by the pressure of the sea ice and explorers were lost in the darkness and cold. One of the biggest stepping stones in the discovery of the Arctic was the tour of *Nordenskiöld* in 1878-1879, where he was able to pass along the whole *coastline of Siberia* from Norway to the Bering Strait. Another Scandinavian, *Amundsen*, was the first to traverse

Eric the Red discovered Greenland in 982 and Leif Eriksson North America around year 1000.

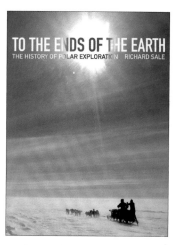

Books on the story of exploration of the Polar regions are highly interesting.

Nordenkiold discovered the North-west Passage and Nansen the North-east Passage.

the western passage from Greenland through the Canadian Archipelago to Alaska in 1905. In 1909 the American *Peary* claimed he had reached the North Pole on foot. Even though he probably did not reach the Pole, his attempt caught public attention. Since then various people have reached the pole on skis and on foot, including the Icelander Haraldur Ólafsson. Part of the difficulty of accomplishing this feat is that the ice is propelled by the gyre and often moves up to 6 or 10 meters each day. Modern GPS technology assures that those who claim to have reached the Pole actually have done so.

Because the *Arctic sea ice* is floating in the Arctic Ocean, it was possible with the advent of nuclear submarines to traverse the Arctic Ocean under the ice. In 1959 the first nuclear submarine, the *USS Skate,* reached the North Pole and was able to surface by finding thin ice to break through. In the 1960's there was a large increase in the building of icebreakers, many of them powered by nuclear energy. *Remote sensing* and *icebreakers* combined mean that the Arctic Ocean is navigable. The first icebreaker to reach the North Pole in this way was the USSR *Arktika*, in 1977. The Soviets, and today the Russians, operate many icebreakers in the Arctic, mostly for breaking up channels where

cargo ships and oil tankers can follow. This makes ocean transportation possible along the Siberian Coast. Of late some Russian icebreakers, including the nuclear-powered *Yamal*, have started to be operated as tourist ships in the summertime, even providing regular trips to the Pole.

Because of the *Cold War* and the hostility between the *Soviet Union* and *the West*, there was very little co-operation on scientific research and other matters concerning the Arctic. In the 1980's, however, as the Cold War started to thaw, a number of joint research projects and co-operative efforts started among the eight nations surrounding the Arctic Ocean. However it was not until 1996 that The *Arctic Council* (AC) was established to include Canada, Greenland (Denmark), Finland, Iceland, Norway, Russia, Sweden and the USA. The governmental status of this council is still somewhat unclear, but in addition to the representatives of the eight nations, indigenous peoples' organizations are included as *government participants*. Also, several countries and organizations have the status of *government observers*. Several programs and institutions that had been established earlier have now been organized under the helm of the Arctic Council. Following is a brief review of the six most important of these.

Because the Arctic ice floats on the sea, it is possible to navigate the Arctic Ocean in submarines under the ice! The first submarine to surface at the North Pole, by breaking through thin ice, was the USS Skate in 1959. The cover of the ice may still have military importance throughout this century.

The first we should mention is *The Arctic Council Action Plan* (ACAP). The main objective of this action plan is to work against pollution of the Arctic. The ACAP has several guiding principles, notably the recognition and use of traditional knowledge, the need to co-operate on a regional basis for protection and preservation, and the integration of Arctic environmental concerns in economic administrative and research sectors. In order to supply the researchers and policymakers with the necessary tools in terms of maps and monitory devices, another institute was established: *The Arctic Monitoring and Assessment Program* (AMAP). The third institute to be mentioned here is the *Conservation of Arctic Flora and Fauna* (CAFF). This is a distinct forum of Arctic professionals, indigenous peoples' representatives, and observer countries and organizations established to discuss and address circumpolar Arctic conservation issues. The fourth working group is *Emergency Prevention, Preparedness and Response* (EPPR), which exchanges information on best practices and conducts projects, for instance on the development of guidance and risk assessment methodologies as well as response exercises and training. The work has focused mainly on oil and gas transportation and extraction, and on radio-logical and other hazards. In 2004, the EPPR was directed by the Arctic ministers to expand its mandate to include *natural disasters*.

Extended use of *natural resources* (oil, gas and mining), and *growth in tourism* will lead to new and more frequently used navigation routes. This calls for new efforts to enhance the security of marine transport, preventing emergencies or responding to them efficiently, including cross-border assistance among neighbouring states. The fifth program deals with *The Protection of the Arctic Marine Environment* (PAME). It was established in 1993 as one of four programs of The Arctic Environmental Protection Strategy (AEPS). These programs, in addition to a new *Working Group on Sustainable Development* (WGSD), now operate under the auspices of the Arctic Council.

In addition to these six main institutions or working groups of the Arctic Council there exist also several other groups or organizations that are involved with matters of the Arctic. In the following, some of them will be described briefly (see also glossary on page 164). The first organization to be mentioned is T*he International Arctic Science Committee (IASC)*. This is a non-governmental organization whose aim is to encourage and facilitate co-operation in all aspects of Arctic research,

Logos of four of the Arctic Council programs listed in the organizational chart to the left.

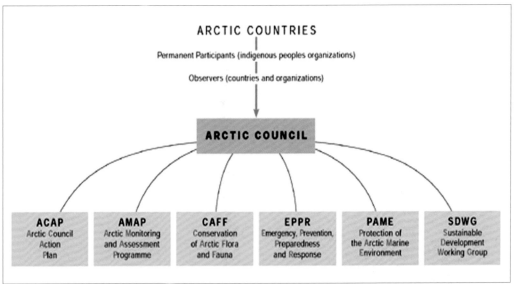

The organizational chart of the Arctic Counil that is composed of the Arctic countries, permanent participiants, and also has several observers. The six main operational and research programs of the Arctic Council are listed in the lowest row and are described in the text on this spread.

The International Polar Year: An important event that puts Polar matters into the limelight.

in all countries engaged in Arctic research and in all areas of the Arctic region.

The IASC member organizations are national science organizations covering all fields of Arctic research. An international science program plan recommended by IASC should be of high priority for Arctic and global science. A very important international project of *The Arctic Council* and the *IASC* is *The Arctic Climate Impact Assessment* (ACIA). This study was founded to evaluate and synthesize knowledge on climate variability, climate change, and increased ultraviolet radiation and the consequences. The results of the assessment were released at the *ACIA International Scientific Symposium* held in Reykjavík, Iceland, in November 2004. This report is based on the work of around 300 scientists and was published in a book of more than 1000 pages in 2005. A special report was also written on the human dimensions of the Arctic: *The Arctic Human Development Report* (AHDR).

Two conferences on Arctic research have been conducted of late. The second was called *The Second International Conference for Arctic Research Planning* (ICARP II). This conference was held in Copenhagen, Denmark, in November 2005. It had the task of helping to create Arctic research plans, plans that are meant to guide international co-operation over the next 10 to 15 years.

The largest upcoming event in the discussion of Polar matters will be *The International Polar Year* (IPY), which will take place from March 2007 to March 2008. The main objectives of the *Polar Year* are to explore new scientific frontiers, to increase our ability to detect changes, and to capture the interest of schoolchildren, the public, and decision makers.

An important scientific organization, *The Arctic Ocean Sciences Board* (AOSB), was established in 1984 to fill a recognized need to co-ordinate the priorities and programs of countries and institutions engaged in research in the Arctic Ocean and adjacent areas. The long-term mission of the AOSB is to facilitate Arctic Ocean research through the support of multinational and multidisciplinary programs in the natural sciences and various fields of engineering.

Another institution involved in oceanic research is *The Arctic and Sub-Arctic Ocean Flux* (ASOF) program. The ASOF aims to measure and model the variability of fluxes between the Arctic Ocean and the Atlantic Ocean, with the view of implementing a longer-term system of critical measurements needed to understand the high latitude ocean's steering role in decadal climate variability.

In 1995, the *European Polar Board* (EPB) was established. The EPB is *The European Science*

The Arctic Human Development Report, the first overview of the human dimensions.

Languages of the Arctic. Blue: Eskimo–Aleut, reddish: Na-Dene, greenish: Chukotko and isolated languages, brownish: Altaic, yellow: Uralic.

Military sites (blue squares) often contain much contamination. The Arctic was a distant area and polluting activities were often located there.

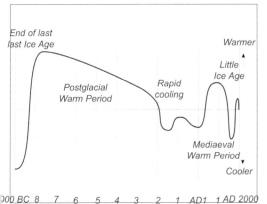

In the Medieval Warm Period Indians in North American moved northward. In Europe the Norse sailed west to settle Iceland and Greenland.

Foundation's expert committee on *science policy in the Polar Regions*. It acts as a voice and a facilitator of co-operation between European national funding agencies, national Polar institutes and research organizations. It also offers independent strategic advice on science policy in the Polar Regions of the European Union, national governments and international Polar bodies.

A more informal research unit is *The Forum of Arctic Research Operators* (FARO) that aims to encourage, facilitate and optimize logistics, and

transport operational support for scientific research in the Arctic through international collaboration for all those involved in Arctic research. An initial meeting was held in August 1998 attended by 24 operators of 11 countries.

Many university departments, institutes and programs have also been established of late. *The University of the Arctic* (UArctic) was established in 2000. UArctic is different from a typical university. It does not have a main campus with buildings or residences. Instead, UArctic is a network of universities, colleges, research institutes, and other kinds of organizations which are located across the *Circumpolar North*.

The *Arctic Monitoring and Assessment Program* (AMAP) listed above has the very important function of making maps of the Arctic and sub-Arctic areas. Previous mapping of this area was not sufficiently accurate as most maps are based on the *Mercator projection* that shows the world as a ribbon, with the Arctic area cut off or impossibly widened. In addition, the research and cartographic foundations required for compiling such a database of the Arctic have been largely missing until the last few decades. The AMAP has a website www.amap.no where an interactive *Arctic map program* can be accessed. This file has several map layers and the viewer can compose his own indi-

Logo of The University of the Arctic symbolizing the flowering of co-operarion on Arctic concerns.

- *Age accumulation of Hg in liver of polar bears*

- *Age structure of Alaska indigenous peoples*

- *Average annual UV (Ultra Violet Radiation) dose*

- *Collective dose rates from European nuclear fuel reprocessing plants*

- *Deposition of PCBs*

- *Geometric mean concentrations of mercury, lead and cadmium*

- *History of temperature changes in central Greenland over the last 40,000 years*

The yellow signs show the sites of dumped radioactive waste and nuclear reactors in the Barents Region. T indicates nuclear weapons test site.

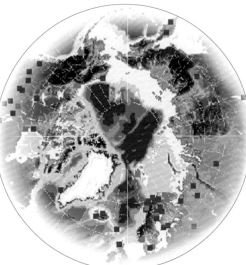

This map shows the locaton of the university and research institutes that are members of the cooperative unit the University of the Arctic.

Examples from the Maps & Graphics Database of the AMAP institute in Norway.

The husky, a hardy animal, is trained to pull sleds over snow and ice in the Arctic.

vidual theme map of the Arctic by selecting desired layers and then printing the map. This map tool is very good for everybody interested in the Arctic. The lack of information on the Arctic is probably most evidenced by the fact that the *first encyclopedia* ever of Arctic matters was published in 2005. This is a huge compilation of more than 2000 pages in three volumes. It includes many of the AMAP maps, as well as several other types of illustrations. Examples of entries in this encyclopaedia are: *Alaska Native Claims Settlement Act; Currents in the Arctic Ocean; Oil Exploration;* and *Ozone Depletion.*

It was not until around 1990, with the collapse of the *East Bloc* and the *Soviet Union* – together with the realization that the global climate was warming – that nations started in earnest to realize that the cold and forbidding areas of the Arctic could become very important in terms of global navigation and resources in the future. The growing awareness of the huge potentials and riches of the Arctic has understandably of late *intensified border conflicts* among the Arctic nations.

The issue of *economic* and *fisheries zones* as well as of *legal borders* is a very complicated subject, with different lines needing to be drawn on maps to define the extent of these three fields. Additionally, certain places in the Arctic are of huge importance, not least islands, because even if the islands are only skerries or very small, they are *outposts for the baselines* of the two-hundred mile economic zones of the countries in question. In the Nares Strait between Ellesmere Island in Canada and Greenland there is a very small but

yet stratically important island called *Hans Island.* This island, which was claimed by early Danish explorers, has been shown as belonging to Greenland on maps.

The Danish explorers are said to have left a bottle of whisky on the island with a note attached, offering the next visitor the chance to drink a toast to their discovery. However, this island, situated as it is in the middle of a very narrow strait only 40 km wide, where major shipping lanes will pass in the future, is no joking matter. The *Canadians* have, for a long time, also stated their claim to this island, and to follow up on their claim, they sent a warship there in the summer of 2006 to secure their possession of it.

As we have now seen in this chapter, there is a *logical progression* from exploration to scientific research and to the founding of international organizations. All this development has, more or less, taken place in only about 150 years, with most of the scientific and organizational activity only taken place in about the last 15-20 years.

It is a strange feeling that until very recently *this future frontier of the world* was *almost totally unknown to the majority of mankind,* but as has been outlined in this book, with global warming – especially if it gets excessive – precisely this area will become one of the most important in the future.

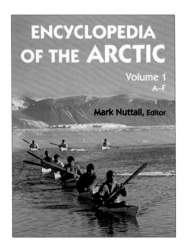

The first encyclopedia ever on the Arctic, was published in three volumes in 2005.

Stefansson, an Icelandic Canadian, emphasized in his book "The Friendly Arctic" that if people adjust to Arctic conditions living is easy there.

Akureyri in Iceland, the location of The Stefansson Arctic Institute, which directed the writing of the "Arctic Human Development Report".

3 Environment, Resources and Development

Much activity is to be expected in the Arctic in the future because of global warming, mostly the opening of global shipping routes and easier accessibility of huge oil and gas reserves.

In this section, we will review, in more detail than earlier in the book, the environment of the high North and its resources. In the second part of this section, we will also gain insight into what kind of developments these resources and improved climatic conditions have already made possible. Until recently estimations of global reserves of oil, gas, and minerals hardly included the Arctic because the area was largely covered with snow and ice for the greater part of the year. Discovery of the scope of mineral reserves and possibilities for development in the Arctic is just beginning. The continental shelf under the Arctic Ocean is relatively flat and not deep (see the map to the right). This means that the bottom of the ocean is easily accessible for the exploitation of minerals and oil.

One of the Sub-Arctic's main potentials in the future is in the area of food production. The *development of agriculture* in its vast terrain is just starting and *will proceed at a fast pace* as the warming continues.

Although this book has mostly focused on the opportunities and positive developments that are opening up because of global warming, we should not forget that there will also be very negative impacts, mostly, however, in the already warm areas of the globe. Early in the book a *balanced overview* of the changes that come with global warming *was given in two matrixes* on the spreads on pages 22-23 and 26-27. On page 23, for instance, there is an overview of what warmer temperatures mean for the cold areas of the globe.

In general it can be said that with increased warming of the North, productivity of all its life systems will increase. A crucial fact is that more precipitation will occur in the Arctic as it gets warmer, contrary to what will happen in many

southern areas. With warming, vegetation in the North will actually become more "southern". Therefore the area will yield *more and more valuable agricultural products* in the future. The increased warmth of the Arctic Ocean and increased amount of nutrients that will be discharged by the increasingly ice-free Arctic rivers will mean that the *Arctic Ocean will take a big leap in marine productivity*.

Today some of the fishing grounds are closed by the Arctic sea ice, but as the warming continues the sea ice will continue to retreat. Some predictions say that as soon as 2100 the Arctic Ocean will be totally ice free in the summer. The fact that large areas of the Arctic Ocean are shallow means, in addition to the retreating sea ice, that more sunlight can reach the bottom and make it a fertile ground for fish and other marine organisms.

As fishing needs to be a year round activity to reach maximum profitability, this highest productivity will not be gained until the sea ice has disappeared in the wintertime also. This will not

One of the greatest possibilities of peoples in the Arctic is more fishing as the sea ice retreats further.

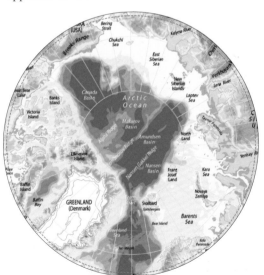

The depth of the Arctic Ocean is characterized by the contrast between its very shallow waters (light blue) and its very deep waters (dark blue).

0 m
−100 m
−500 m
−1,000 m
−2,000 m
−3,000 m
−5,000 m
Greatest depths

Glacier

Barren land

Mountain Tundra

Lowland Tundra

Northern Boreal Forest

Middle Boreal Forest

This map shows the location of vegetation types in the Arctic and the Sub-Arctic. Their location will change very much with global warming, as all vegetation types move north. How much will depend on the rate and the intensity of the warming.

As permafrost thaws, roots of trees lose their hold and the trees start to lean. Called drunken forests.

happen until sometime in the 22nd century. Then the Arctic Ocean will probably become one of the richest fishing areas of the world.

The map above shows the distribution of vegetation of the Arctic and Sub-Arctic areas. The character of the vegetation, or its absence, is a direct indicator of the quality inherent in the land. Therefore, the brown areas on this map, which indicate *barren land*, are obviously – in addition to the *areas covered with ice*, shown in lilac grey – the areas least suitable for growth and therefore also

lest suitable for human habitation. The *lowland tundra areas*, shown in bluish grey, are mostly located along the coasts. With the warming of the climate the *boreal forests*, pale green on the map, will extend over most of these areas.

With the northern drift of the northern boreal forests, the *middle boreal forests*, middle green on the map, will extend further north. The dark blue-greyish colour on the map shows *mountain tundra*. Here the vegetation is very scarce because of the cold climate at the high altitudes. These dark coloured areas, therefore, are those areas of the high North – in addition to the barren and ice areas – that are least *suitable for human habitation* or even *exploitation of minerals* because of the difficult climatic conditions.

In 2002, a publication called GEO3 *Global Environment Outlook* appeared. This publication contains an estimation of *human impact in the northern areas of the globe* as of now, as well as an estimation of how much these human impacts on the North will have been *extended in the year 2032*. The two maps shown below are a modified version by the *AMAP* institute in Norway. We see by comparing the two maps that in the next 30 years this report expects a vast extension of human activity to the North.

In the book *Vital Arctic Graphics*, where these modified versions of the maps are published, it is emphasized that the 2032 picture is a *rather conservative estimate* of what will happen in the North in the next decades. The book points out, for instance, that GEO3 underestimates the rapid oil exploration and development activities taking place on the *North Slope of Alaska* and in the *Mackenzie Valley region in Canada*. The book also points out that in the estimation of UNEP 2001, which is a publication of the *United Nations Environmental Programme*, it is projected that perhaps as much as *80% of the Arctic land area* will become *impacted by development by 2050*,

if current trends continue.

The perspective of this present book is directed still further into the future, mostly to the year 2100 because that is the reference year for most of the predictions in the *Arctic Environmental Impact Assessment Report* (ACIA). The present book, however, in some instances looks still further into the future, into the 22nd century. The author of this book has already, in previous chapters, shown in words and maps, what in his opinion the warming will lead to in terms of development in the Arctic – especially if the warming continues beyond 2100.

It has been argued in this book that with excessive future warming, the North will be practically the only area that mankind can comfortably escape to from areas of intolerable heat because by then the Arctic will have a comfortable, or slightly cold, climate with vegetation somewhat similar to what we see in Scandinavia today. The loss of the tranquil Arctic environments, glaciers and tundra will be sad, but global warming is coming and we *should not paint the changes in nature* that come with it as *only bleak and negative*. It is, rather, advisable for us *to embrace the inescapable changes* with *a positive attitude* and to start to discover how we can adjust to the upcoming changes in the world. The author goes as far as to say that the warming

The logo of a project of the Arctic Council to assess the impact of a warming Arctic.

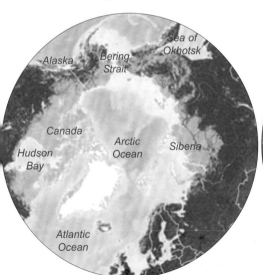

These two pictures show, in brownish colours, an estimation of the intensity of human impact in the North. This picture shows the range in 2002.

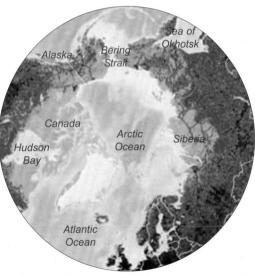

Estimation of human impact in the North by 2032. This is a conservative estimate that, e.g., does not include some large oil extraction projects.

of the globe's climate could possibly lead to *a more productive and positive natural environment than we have today*. Warming in general and more precipitation means more productivity of the natural biosystems, and the very cold polar areas of the globe that until now have been almost totally unproductive in terms of maintaining human life will become free from the bondage of ice and cold. There are of course areas in the middle section of the globe that will be hit hard by the warming, not least because in these areas precipitation will decrease and evaporation increase, leading to the spreading of deserts and lower yields from cultivated land. The positive attitude promoted here even claims that *a warm Arctic is a new Paradise*, and a place to escape to from the possible excessive heat of the future. We have already seen a preview of this as central and southern Europeans flock in the summer to Scandinavia to escape the intolerable heat spells.

In the past there have been many settlements and areas of the world that we have not treated with respect. The result, in many places, has been huge environmental damage, and adding the depletion of resources that has resulted from thousands of years of habitation in some of the central areas of the globe, we see that the Polar areas and their resources, and their future benign climates, are really *a gift to humanity*. However, we need to be very aware, especially today and in the next few decades, that there will be, as in most other areas of the globe, huge changes in the ecological systems of the high North. This will make

The lack of warmth is the main reason why the high North is so barren. Therefore global warming will increase growth there, both on land and in the oceans.

The Exxon Valdez accident at the Alaskan coast in 1988 showed people how serious the impact of oil spills can be on the living environment.

the Arctic environment still more fragile than it is today. Eventually *a new balance in the ecosystems will be reached* that will be able to withstand more stresses of development. However, in this century, we need to be very aware of the fragility of the Northern ecosystems as we increasingly intrude on the area. Therefore *very stringent planning processes need to be installed* for every major project of the Arctic, and that is the subject of the next and the last section of this book.

As described before, huge activity in many Arctic areas has already started, mostly in terms of mining as well as oil and gas extraction. The big map to the right shows these areas. The black squares show *the main mining sites*, the lilac areas *prospective areas for oil and gas*, and the red and green spots *areas of today's oil and gas production*. Unbroken purple lines show the *main existing oil and gas pipes*, whilst the broken lines are *projected oil and gas pipelines*. The red and yellow stars at Ekofisk in Alaska and Usinsk in Western Siberia remind us of the huge threat of oil spills that comes with extraction. In these accidents more than 50,000 tons of oil were spilled into the environment. The impact on wildlife, for instance birds, has been terrible, but because oil is a natural product it eventually, over a long period, is absorbed by natural processes, but the damage to life and ecosystems can linger for decades. In addition to the lilac spaces, the map also contains green squares showing *areas of oil exploration* that are outside today's main oil areas. Two of these areas are in the West Coast of Greenland and exploration there has already confirmed that in this area there is probably more oil and gas than was in the whole North Sea before extraction started. On the upper part of the map, close to Alaska and the Bering Strait, we see four more such squares outside the lilac areas that geologists consider the main oilfields of the future.

On page 139 we have already seen a map that presents an estimation of areas of human impact in the high North in 2032. In the earlier text about this map, we mentioned that the perspective of this book goes at least 100 years further into the future. In addition, this 2032 projection assumes that the main infrastructures will be the same. If we assume that the warming will become

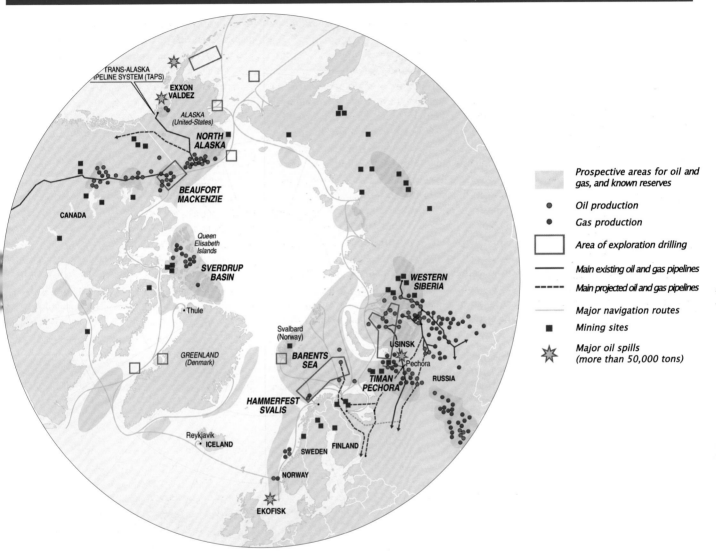

Prospective areas for oil and gas, and known reserves

● Oil production

● Gas production

▭ Area of exploration drilling

—— Main existing oil and gas pipelines

----- Main projected oil and gas pipelines

—— Major navigation routes

■ Mining sites

✦ Major oil spills (more than 50,000 tons)

Industrial development in the Arctic and the Sub–Arctic. Areas coloured with lilac are prospective areas for oil and gas. As of now most such activity is taking place in the Barents Sea, in the Mackenzie Vally in Canada and in the Alaskan North Slope.

extensive, even beyond 2100, we can be certain that all the resources and the possibilities offered by less snow and a better climate will lead to extensive developments in new transportation infrastructures. A projection on how they might develop was presented earlier in the chapter *Transportation Structures* on page 79. On page 88, the author predicts that this will not only be directed to the southeast, to China, but also to the Sea of Okhotsk by Russia's north-eastern coast. As this sea becomes ice free, the sea itself and the surrounding area will offer huge possibilities in terms of fishing, farming and mineral exploitation. The author therefore predicts that a major traffic corridor will develop from there to Central Asia and Western Europe. On page 95 it was explained in somewhat more detail how the settlement and transportation structure of Siberia could possibly develop in the future, leading to an overall picture of what might become the locations of the main urban centres of Central Asia and Siberia in the distant future. After becoming free of ice, the intersections of the suggested main railway lines and the largest rivers will become very important development axes and transportation corridors.

By keeping this picture in mind, we see that it is impossible to make assumptions about what the future may hold, that for now it suffices to extend the old forms of transportation and urban infrastructures. The infrastructures of the future may be fundamentally different from what we see on today's maps. Nevertheless, it is useful to look at the map on the bottom of this page, a map that shows two main *railway corridors* from Northern Europe to China and the Pacific Ocean.

In this picture, a northern railway corridor to the sea of Okhotsk is not shown. This railway project, that is already underway, is one of the largest infrastructure investments of the world. Its main aim is to create a link between China and the Atlantic coast of Eurasia. The very far northern town of Narvik was chosen as the main west-

ern port for both railway lines, and the number of days the journey takes is shown in yellow, starting from Narvik: The train journey to Archangelsk takes one and a half days, to Ürümchi seven days, and the journey to Vladivostok ten to eleven days. The main idea behind this continental transportation link is not to have to go south of Africa with large containerships to reach Western Europe and North America.

With the new railway, the containers from, for instance China, will be transported by train to Narvik and then transported onwards to Europe and the East Coast of North America by relatively short shipping routes. The usefulness of these railways, however, will possibly reach their prime in the earlier part of this century because as shipping routes over the Arctic Ocean open over most of the year, they will provide a much cheaper way to transport goods from the East Coast of Asia to the Atlantic.

Narvik in northern Norway is going to the main Atlantic port of the N.E.W. Siberian railway system.

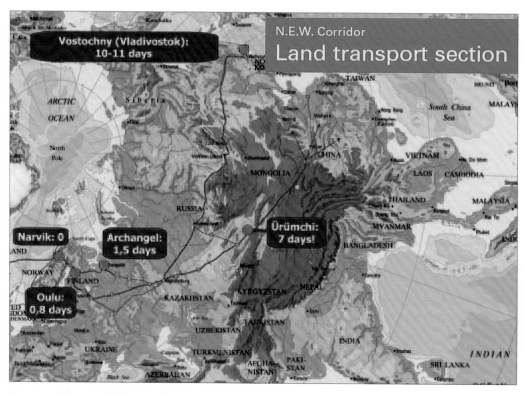

A new and improved Siberian railway project to connect the South-east Asian and North Atlantic markets is now underway. Its importance, however, will decrease as all-year shipping lanes have opened through the Arctic Ocean in the latter half of the 21st century, especially in areas close to ports.

4 Planning Processes in the Arctic

As this book has described, the high North and the Arctic are very likely to become one of the most important *development areas* in the world in the future. The areas are very rich in resources, a plus that is already leading to considerable development. The lessened ice on rivers and the opening of the Arctic Sea routes between the Arctic and Pacific Oceans will in due time make more and more of these resources accessible. A substantial increase of ship traffic along the Siberian coast, for instance, has already opened the coastal areas to increases in development. This future perspective calls for *mixed emotions* because we have, in the last fifty years, come to know the serenity and beauty of Arctic terrains, not least from photographs and films.

The Polar regions are the last areas on the globe to be developed, and in a way, they can be said to have been conserved by the cold and ice. Because of the cold, the Arctic ecosystems are *unique* and at the same time *sensitive,* both to the impact of human activity but especially to stresses that the very rapidly changing climatic features are putting on them. All this calls for *very considerate planning,* followed by *cautious execution* of building projects, industrial operations and shipping.

In recent last decades, important tools have been developed to help us plan in such a way that environmental impacts are kept at a minimum. The main method to prepare and investigate major projects is the *Environmental Impact Assessment* (EIA) methodology. This methodology was first developed in the USA states of California and Washington. In the 1980's the European Union ordered all its member states to adopt this methodology in order to be able to assess in advance what the eventual environmental impact of large projects would be. Based on the findings, authorities decide whether to allow or prohibit the proposed construction or project. In some cases the project planning and design are allowed to advance if certain precautionary measures are taken.

Today's EIA assessments not only consider the impacts on natural and physical environments but also the impacts on *economic* and *social environments.* In the 1990's the EU extended the EIA methodology to a wider sphere, a procedure called *Strategic Environmental Assessment* (SEA). This method is meant to take the evaluation of possible impacts to the higher levels of policy, law, concepts and programs. This is necessary because individual projects are, in all the main aspects, formulated in policies and programs at a high level. As we come to the level of an individual project there is therefore usually no possibility of turning back and stopping, even if the project is not environmentally acceptable. Then most often the only option is to modify the project slightly.

By 1989, *as the Soviet Union collapsed,* people in the West had started to realize that the huge mineral and oil treasures of Siberia were soon to be more readily available. This would be *for three reasons:* the thawing of the political climate, improved ice surveillance techniques, and the retreat of the sea ice at the Siberian coast. This would mean that the *Northern Sea Route* (NSR) would, even within a few decades, be likely to be open all the way to the Pacific. These new developments made international organizations very interested in contacting Russian authorities and scientists about the possibility of huge undertakings in Siberia in the future, undertakings that certainly would need large foreign investment. In 1993 the *International Northern Sea Route Programme* (INSROP) was established, leading to meetings of scientists from various nations. At the first meeting in Oslo in November 1993, one workshop had the task of screening and creating a focus on the relative importance of the marine resources in the NSR area. This was then taken a step further to the selection of value issues, named the *Valued Ecosystem Components* (VEC). This selection was based on the potential or ability of various animal species to interfere with NSR activities.

The third and largest symposium on the

The Arctic is very sensitive. Ordered planning processes can do much in reduce the impact of human activity on nature.

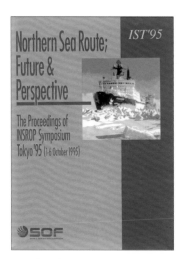

The first large symposium on the Northern Sea Route was held in Tokyo in 1995.

Northern Sea Route was held in Tokyo in 1995. By then scientists had already outlined the main components of the work on environmental assessment and planning that needed to be developed in order to be able to implement Environmental Impact Assessment practices in these little known and little mapped northern regions.

One of the most important components in this process is the Dynamic Environmental Atlas. The creation of such an atlas was absolutely necessary for those conducting the assessments to enable them to know what degree of impact the various types of environments can take in terms of large projects.

The work on the creation of the atlas is organized in five separate but interrelated projects that groups of scientists have been working on for a long time: 1) Invertebrates and Fish, 2) Marine Birds, 3) Marine Mammals, 4) Coastal Zone and 5) Large Rivers, Estuaries and Deltas. This is a continuing task, but the main objective of the project is to establish a database containing information on the distribution, abundance, migration, and breeding and feeding areas for the most important marine animal resources in the area of the Northern Sea Route.

The project also has the proclaimed goal of instigating discussion on the possible impacts of other human activity on these species. The data from the projects is already being used as input for Environmental Impact Assessment (EIA) projects concerning industrial facilities, shipping and navigation activities along the NSR. In such a work it is of primary importance to identify a number of priority issues to be addressed in the EIA investigation.

Although these are the basic impacts to be studied, i.e. impacts on organisms and ecosystems, the impact on activities and projects related to the NSR in terms of the indigenous and other people living in the Arctic, as well as the impact on economic structures of the societies of the areas in question, need to be a part of the evaluation program.

As we see in the scheme to the right, the data from the Environmental Atlas feeds into a computerized Geographical Information System (GIS) that, as the arrows show, also gets input from other sources. As this computer system is estab-

lished, evaluation of the environmental safety of navigation can be launched. As a second phase, in this evaluation scheme, environmental assessment projects are being worked out, leading to, as the lowest line on the scheme shows: Recommendations, Decision Making and an Environmental Pilot.

The EIA process requires that several different alternatives or scenarios as to how the project in question can be developed. Each of these scenarios is evaluated and compared in terms of impact on the environment. The idea behind this is that the comparison will help decision makers to select scenarios that are likely to be least harmful, but always in relation to the potential gains.

The scenarios that need to be created and described in detail for the EIA fall into several categories. These can be scenarios of types of cargo, scenarios of operational or accidental occurrences, or scenarios that describe the temporal-spatial dimension. The temporal dimension is especially important considering the very dynamic and changing environmental conditions of the Arctic. In addition, there can be certain circumstances that have to be included as a study of special importance. These can include high-risk areas for potential accidents like causing an oil spill from a damaged tanker. This type of scenario usually includes a risk assess-

---

The Dynamic Environmental Atlas

1) Invertebrates and Fish
2) Marine Birds
3) Marine Mammals
4) Coastal Zone
5) Large Rivers, Estuaries and Deltas.

An important Arctic marine transport workshop was held in Cambridge UK in 2004.

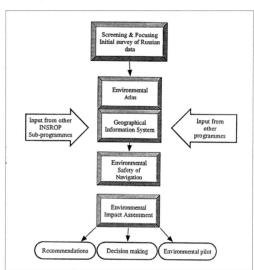

The scheme developed to assess the enviromental impact of the Northern Sea Route. It starts by screening and ends in recommendations.

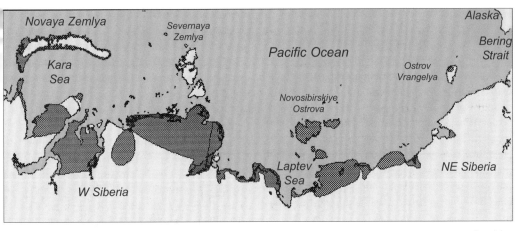

The map shows the main areas of the king eider. An important part of many ecosystems in the Arctic. It is a bird that is an important food source for several animals, and humans too. Its meat, eggs and down are much sought after.

The male common eider is spectacularly coloured, but the female is drab as she sits on the nest.

ment that takes into account the possible risks, relative to the size of ships and cargo and the areas possibly affected. The special and unusual problem in conducting an environmental impact assessment in the Arctic is the very *rapid rate of change*. For instance, it does not make much sense to describe the impact on *walrus colonies* in a certain area if, because of warming climate, the walrus are going to move away from the area to other more remote and colder areas.

Another area of great difficulty in terms of assessment is the *socio-economic impact* of the projects in question. To maintain the present socio-economic structures in various areas in the Arctic may not be possible in the face of upcoming occupational and climatic changes. Social upheaval, and changes in general, most often mean difficulties, but the abundance of jobs that become available for both indigenous and other people living in the Arctic means that income and standard of living will improve. These are the gains, but global warming and the activities that are going to take place in the Arctic are unfortunately bound to mean the end of many of the ways of the *traditional societies*. Many projects are fortunately already in operation to maintain the knowledge and practice of traditional activities, even as they watch television and cash their gov-

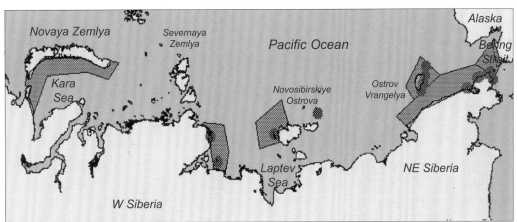

The picture shows walrus colonies at the Siberian coast. Walrus are sensitive to human activity so, if it is possible, all major shipping traffic and the building of large-scale industries and human settlements should be avoided in the grey areas on the map.

Walrus are partly dependent on ice, so retreat of the ice means that walrus will try to move north.

ernment social security checks.

Warming will also drastically change the *ecosystem patterns* of the North. The life of the polar bear and the migration patterns of the caribou, for instance, will be affected in ways that are difficult to project. The sea ice is "land" for some species like snowy owls, gulls, walrus and polar bears; its removal will inevitably force ecosystem changes. Judging from earlier periods of extensive polar warming, like that of 4000 and 8000 years ago, these species will likely be able to rely on other ways to survive and feed. Walrus and seals haul out on land to rest, and polar bears can fish in rivers, like their Kodiak Alaska cousins.

The *Arctic Climate Impact Assessment* (ACIA) report for the whole Arctic region, written by more than 300 scientists from 2002-2004, will prove to be very helpful in assessing all types of small-scale projects within the Arctic in the future.

The Arctic Council also had another report written in the same period, the *Arctic Human Development Report* (AHDR). This human development report is the first comprehensive attempt to document and compare systematically the welfare of Arctic residents on a circumpolar basis. It seeks to expand our horizons by spotlighting the socio-economic and cultural aspects of the lives of the people in the region. This report is actually an integral part of the evolution of regional co-operation in the Arctic, involving, for instance, *Arctic parliamentarians*. This special project arose because of difficulties in devising a coherent agenda for the Arctic Council's *Sustainable Development Program* (SDP).

The two main components of this report are the *Core Systems* and the *Crosscutting Themes*. The Core Systems' component includes: *Societies and Cultures, Economic Systems, Political Systems*, and *Legal Systems*. The Crosscutting Themes are: *Resource Government, Community Viability, Human Health and Wellbeing, Education, Gender Issues and Circumpolar International Relations* and *Geopolitics*.

Much of the mapping material for this report was produced by the *Arctic Mapping and Assessment Program* (AMAP) in Arendal, Norway. In 2005, that institute published the first book of *Arctic Graphics and Maps* under the aegis of several institutions.

Three maps on this spread are taken from this book. The first map to the left below gives the *population distribution for the Arctic*. It comes somewhat as a surprise how small a part of the population in most of the countries is indigenous, as

The life styles of indigenous and other people are changing fast because of the introduction of modern ways, and because of the warming.

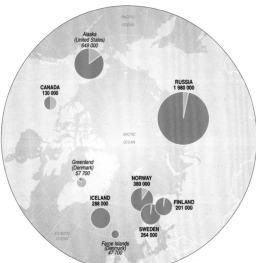

Population numbers in the Arctic. Yellow shows the share of indigenous people. Except for Canada and Greenland they are in the minority.

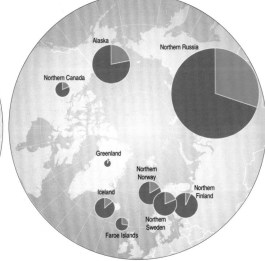

The size of the economies in the Arctic in 2002, for Alaska e.g. USD in PPP 30,000 million. Green: Primary sectors (oil...). Blue: Other.

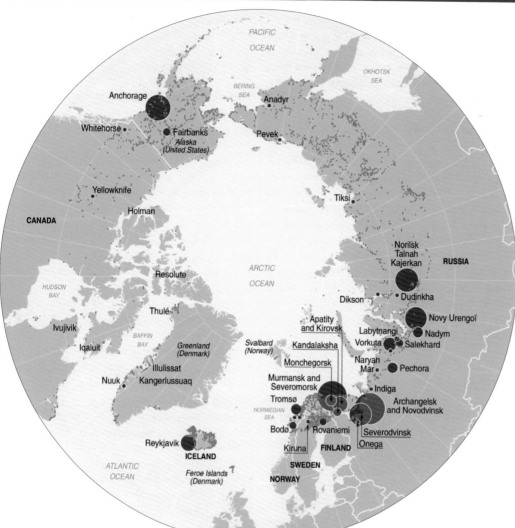

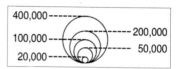

The sizes of the circles indicate the size of populations in the Arctic.

The map shows today´s indigenous settlements in the Arctic. As in the past, today´s settlements are usually located in areas that fit well for caribou (reindeer) or hunting of sea mammals. An increasing number of people do not depend on this but work in more modern professions.

shown in yellow within the blue circles. In the map to the right the circles indicate the *size of the economies* in the countries surrounding the Arctic Ocean.

The gross product is measured in terms of millions of US dollars as *Purchasing Power Parity* (PPP). It comes somewhat as a surprise that the *primary sectors* of fisheries, oil, mining and agriculture, shown in green within the blue circles, constitute only a small part of the gross product. The primary sectors of Alaska and Russia, however,

loom larger as in these areas oil and mining are of such magnitude. These two sectors are bound to expand hugely in the future. In the Northern European countries and Canada, the primary sectors in the Arctic are mostly fishing and agriculture, though constituting, only about 30% of the gross product.

Of course, social studies of the Arctic and Sub-Arctic regions focus mostly on the traditional indigenous settlements. The large map above shows the *location and size of these settlements*. The

A small Eskimo settlement in Greenland. Gradually larger settlements have started to develop.

The polar bear lives and hunts mostly on ice. The disappearance of the ice, however, in earlier periods did not mean its extinction.

A New World

*The Arctic from above
Is like a new planet.
Frost and isolation
Made it unknown.*

*A warm Arctic,
Is a new paradise;
A place to escape to,
From excessive heat.*

*Exploration and settlement,
Once a pride –
Have soured
From bad conduct.*

*Arctic settlement
Has to be planned
To make sure
It is sustained.*

small blue spots represent villages with less than 20,000 inhabitants as well as very small communities. As we see, the original settlement structure is very spread out, mostly because these were people who hunted over a large area and have had to adjust their terrain of habitation to attainable resources.

Because many of these communities live from sea mammals and, in later times, increasingly from fishing, many of these settlements are located on coasts and fjords as well as on estuaries of large rivers. It is apparent from the map that the largest indigenous settlements are in Northern Scandinavia, on the Kola Peninsula, close to Archangelsk, and in Western Siberia in general. It is exactly in these same regions where most of the oil reserves have been discovered. This uncomfortable fact will unavoidably *lead to more conflicts* between traditional lifestyles and the highly modern industrial lifestyles that come with oil extraction and shipping.

We have now, in these last sections of the book, reviewed conditions concerning several sectors and issues of the Arctic. The reader will by now have recognized how complicated the planning and social issues are going to be in the Arctic in the future, certainly leading to confrontations of different opinions on ways to proceed. Most of us will regret that the climate and the nature of the Earth – especially that of the Arctic – are changing so fundamentally. Today most scientists agree that *the climatic changes cannot be stopped.* Everybody, however, agrees that it is *advisable to slow these developments down* as much as possible. This will reduce the troubles and also, in the process, give peoples and ecosystems time to adjust to the changes.

This book has not been written by an expert on the climatic or natural sciences, but by a planner. In the last few decades he has observed how very serious attempts have been made to stop the increased use of fossil fuels and thus the rate of greenhouse gas emissions, which are the main reason for global warming. The whole world has been preoccupied by these attempts, but *scenarios* of what will happen *if the warming cannot be stopped* have been very little discussed, except for negative and doom-like descriptions of the impacts on natural environments.

The stage where a planner usually enters the picture is when climatologists and other scientists have collected all the relevant scientific material for the area in question. As for the Arctic, it is only with the very recent publication of the various reports that have now been described that the *information needed for a planner to start to synthesize from these data* has become available. It therefore became possible only very recently for a planner – such as the author of this book – to create a *scenario* on what is likely to happen in terms of *the settlement and activity structures of the globe in the future.*

To create such a scenario was the main task of this book. In the second half of the book the author has given examples of how settlement structures of individual areas or countries might develop in the future, and this sixth and last chapter of the book has provided more detailed insight into the developments that are already taking place in the Arctic.

As we have seen in this last section, the people, the associations and the governments responsible for how the future will develop in the high North have been undertaking *very important steps* in terms of information gathering, mapping and writing reports.

Environmental impact assessment and planning procedures have also been installed. This work needs to be continued in a very earnest way. If we assure viable and justifiable ways for the settling of *this last frontier on Earth,* many of the previous mistakes of the planning of settlements in the older part of the globe can be avoided. The main tools to help this happen are the principles: *sensitivity to environmental* and *social issues* and the concept of *sustainable development.*

References for the Chapters

This list of references gives specialized documents and websites used in writing this book, ordered by chapters. Each reference starts with the page number it refers to and – if it deals with a direct quote – the column (C) and the line (L) are mentioned. Full lists of *General References* and *Websites* are given on pages 162 and 164.

Introduction (pp. 7-11)

7   *United Nations Framework Convention on Climate Change* (UNFCCC) http://unfccc.int/

7   *Asia-Pacific Partnership on Clean Development and Climate* http://www.sustainibility.act.au/greenhouse

8   *Almanac of Geography,* National Geographic (2005). Map projections, p. 29.
Many atlases describe various types of map projections and their positive and negative characteristics.

12   All 24 verses of the poem are given on p. 168

GLOBAL WARMING
I   *Warming: The Catalyst of Change* (pp. 13-36)

13   *Intergovernmental Panel on Climate Change* (IPCC) www.ipcc.ch

14   *Arctic Climate Impact Assessment* (ACIA) www.acia.uaf.edu Chapter 4

14   www.amap.no/acia

15   *United Nations Framework Convention on Climate Change* (UNFCCC) http://unfccc.int/

16   *Asia-Pacific Partnership on Clean Development and Climate*
http://www.sustainibility.act.au/greenhouse

18   *Arctic Climate Impact Assessment* (ACIA) www.acia.uaf.edu Chapter 1

19   *Intergovernmental Panel on Climate Change* (IPCC) www.ipcc.ch Climate Change 2001

20   *Arctic Climate Impact Assessment* (ACIA) www.acia.uaf.edu P. 26-8

21   *Design with Nature,* Ian McHarg (1969)

21   *Arctic Climate Impact Assessment* (ACIA) www.acia.uaf.edu Chapter 13

24   "Náttúruvá á Íslandi" (Natural Hazards in Iceland), Ragnar Sigbjörnsson and the author of this book, AVS (1/1996)

24   "Managing and interpreting uncertainty for climate change risk", Rodney R. White. Research & Information 32/5 2004.

25   *Arctic Climate Impact Assessment* (ACIA) www.acia.uaf.edu Chapter 16

26   As for p. 24

28   As for p. 21

28   *Symposium at the University of Iceland* in April 2005 on the impacts of rise of sea-level www.hi.is/page/flod

31   *The Skeptical Environmentalist – Measuring the Real State of the World,* Bjørn Lomborg (2005)

32   C1 L6 *The Skeptical Environmentalist – Measuring the Real State of the World,* Bjørn Lomborg (2005), p. 635

33   *United Nations Framework Convention on Climate Change* (UNFCCC) http://unfccc.int/

33   *Asia-Pacific Partnership on Clean Development and Climate* http://www.sustainibility.act.au/greenhouse

33   C2 L5 *Macfarlane quote*:
http://news.bbc.co.uk/1/hi/sci/tech/4602296.stm

33   www.whitehouse.gov/news/releases/2006/01/20060111

33   *Energy by source*
http://encarta.msn.com/media_461518118/World_Energy_Production_by_Source.html

34   http://worldenergyoutlook.org/2006.asp

34   "Organic carbon in Iceland: Andosols – geographical variation and impact of erosion", Hlynur Óskarsson et al., Catena 56 (2004), pp. 225-238. www.elsevier.com

35   *Arctic Climate Impact Assessment* (ACIA) www.acia.uaf.edu Chapters 14 and 17

35   *Global warming 'cannot be stopped'*
http://www.timesonline.co.uk/article/0,,2-2341516,00.html

35   *Can we defuse the Global Warming Time Bomb?*
http://naturalscience.com/ns/articles/01-16/ns_jeh.html

35   *Kyoto 'will not stop global warming'*
http://news.bbc.co.uk/2/hi/uk_news/politics/3131285.stm

PATTERNS
II How Global Patterns Change (pp. 37-58)

37 *Order out of Chaos – Man's new Dialogue with Nature*, Ilya Prigogine and Isabelle Stengers, (1984)

38 *The Structure of Scientific Revolutions*, Thomas S. Kuhn, (1970)

38 *The Way of Life – Lao Tzu*, (1995)

39 *Forecasting planning and Strategies for the 21st Century*, Spyros G. Makridakis (1982), p. 69

40 "Jardskorpuhreyfinar og haed sjavarstodu", Halldór Geirsson. www.hi.is/page/flod

41 *Collapse – How Societies Choose to Fall or Succeed*, Jared Diamond (2005)

42 *Atlas of 20th Century*, Richard Overy (2005)

43 *Climate Change – Turning up the Heat*, Barrie A. Pittock, (2005)

43 *Collapse – How Societies Choose to Fall or Succeed*, Jared Diamond (2005)

43 *Flood casulties.* *http://www.nbc10.com/news/4030540/detail.html*

45 C1 LX *The Skeptical Environmentalist – Measuring the Real State of the World*, Bjørn Lomborg (2005)

45 C1 L7 Ibid., p. 87

45 C1 L35 Ibid., p. 159

45 C2 L2 Ibid., p. 301

47 *Megatrends: Ten new Directions Transforming our Lives*, John Naisbitt (1982)

48 *The City in History*, Lewis Mumford (1974)

48 *Cities in Civilization*, Peter Hall (1999)

52 *North Meets North – Navigation and the Future of the Arctic*, Ministry for Foreign Affairs, Iceland (2006) 53 http://en.wikipedia.org/wiki/Image:Biosphere2

55 *Atlas of 20th Century*, Richard Overy (2005) 56 http://en.wikipedia.org/wiki/Little_Ice_Age

57 *Human migration*, Encyclopedia Britannica http://search.eb.com/eb/article-9041473#154099.hoobe

58 *Immigration to the USA*, Wikipedia http://en.wikipedia.org/wiki/Immigration_to_the_United_States

III How Spatial Systems Change (pp. 59-78)

60 *Buckminster Fuller - An Autobiographical Monologue*/Scenario, Robert Snyder (editor) (1980)

62 *North Meets North – Navigation and the Future of the Arctic*, Ministry for Foreign Affairs, Iceland (2006)

63 *The City in History*, Lewis Mumford (1974)

66 *Arctic Climate Impact Assessment* (ACIA) www.acia.uaf.edu Chapter 16

68 "Northern Sea Route – Future & Perspective". Proceedings of the INSROP Symposium Tokyo '95, Hiromitsu Kitagawa (editor) (1996)

70 *Arctic Marine Transport Workshop 28-30 Sept. 2004.* Lawson Brigham and Ben Ellis (editors). Held at SPRI, Cambridge (2004)

70 *Vital Arctic Graphics.pdf* at www.grida.no

72 *Ice Riding Information for Decision Making in Shipping Operations* (IRIS) http://www.hut.fi/Units/Ship/Research/Iris/Public

74 *GIUK-hlidid*, Gunnar Gunnarsson (1981)

THE FUTURE
IV The Future Structure of the Globe (pp. 79-102)

79 *Europa: Territorial Cohesion* www.europa.eu.int/constitution/en/ptoc44_en.htm

79 *Europa: Regional Policy Inforegio.* www.ec.europa.eu/regional_policy/interreg3

81 *Transatlantic Tunnel* www.law.harvard.edu/alumni/bulletin/2002/summer http://en.wikipedia.org/wiki/Transatlantic_tunnel

84 *North Sea oil* www.ita.doc.gov/td/energy/north_sea_seminar.htm

89 *North Meets North – Navigation and the Future of the Arctic*, Ministry for Foreign Affairs, Iceland

(2006), p. 20 to 41
94 *Vital Arctic Graphics* on www.amap.no
95 www.europa.eu/abc/history
101 As for p. 81
102 As for p. 89

V *Impacts on a Global Scale*
(pp. 103-124)

103 *The Regional Impacts of Climate Change – an Assessment of Vulnerability*, Intergovernmental Panel on Climate Change (1998), p. 10
105 *Arctic Climate Impact Assessment* (ACIA) www.acia.uaf.edu and www.amap.no/acia Chapter 8. Discharges of sediment from rivers.
106 *Watersheds and the shape of Greenland under the ice*. Kalaallit Numaat, Greenland Atlas, p. 31
109 *Gas liquefaction*. www.statoil.com/STATOIL-COM/snohvit/svg02699.nsf
110 *International Hydrological Programme* (incl. maps) http://webworld.unesco.org/water/ihp/db/shiklomanov
112 *Vital Arctic Graphics* on www.grida.no p. 26
115 *The Best that Money Can't Buy*, Jacque Fresco (2002)
116 *Understanding Media*, Marchal McLuhan (1997)
120 *Atlas of 20th Century*, Richard Overy (2005)

VI *The North: The Future Area of the Globe*
(pp. 125-158)

125 *Arctic Climate Impact Assessment* (ACIA) www.acia.uaf.edu Chapters 4 and 6
126 *Asia-Pacific Partnership on Clean Development and Climate* http://www.sustainibility.act.au/greenhouse
127 *Symposium, Marine Research Institute of Iceland* www.hafro.is/symposium
128 *Arctic Climate Impact Assessment* (ACIA) www.acia.uaf.edu. Impacts of a Warming Arctic
131 *To the Ends of the Earth – The History of Polar Exploration*, Richard Sale (2002)
133 www.arctic-council.org
134 *International Polar Year 2007 - '08* (IPY) www.ipy.org

134 *Vital Arctic Graphics* on www.grida.no
135 www.uarctic.org
136 *Encyclopedia of the Arctic I-III*, Mark Nuttall, editor (2005)
137 As for p. 134
139 *Shell Hydrogen* www.shell.com/hydrogen
142 *N.E.W. Corridor: Land Transport Section*. www.uic.asso.fr/
143 "Northern Sea Route – Future & Perspective". Proceedings of the INSROP Symposium,   Hiromitsu Kitagawa (editor) (1996)
144 As for p.143, p. 208
145 Ibid., pp. 218 and 219
146 *Vital Arctic Graphics* on www.grida.no

Maps and Illustrations

About sixtyfive percent of the maps and illustrations have been especially drawn for this book. The author drew them by hand and then they were scanned into a new *Illustrator* drawing program. The special feature used is that this program can make all lines in a scanned picture are "alife". The draftsman, *Karl Jóhann Magnússon,* turned the drafts into the coloured images that appear in this book.

The illustrations are based on a wealth of information taken from statistical tables, historical books, and thematic maps, but mostly they present interpretations of features that related to the book's theme. For the purpose of clarity all the illustrations contain the least amount of information possible. It is a very common fallacy in the design of illustrations to include too much information and far too many details, which only obscure the main meaning or pattern that the pictures are meant to convey. An important part of the conception of the maps and drawings by the author was to find a fitting degree of abstraction.

The author has had *the policy,* for all of his twelve books, of *making the illustrations specially made for these publications available free of charge* for everybody to use, provided that the source is quoted as the illustrations are used. The same permission *applies for all quotes from the texts* by the author in these books.

The illustrations created for this book are the following – citing the page number first, and then the number of the picture:

8-1 How a world map is produced, 8-2 Ribbon map and the poles, 9-3 All directions south, 9-4 Antarctica's system of directions, 10-1 Ribbon cut in three places, 17-1 Northern areas above 200m, 17-2 Habitable areas in the North, 21-1 Shipping route NW Siberia-NE America, 22 and 23 Matrix: Changes with global warming, 25-1 Sequence in changes, 25-2 Warmer-Wetter-Windier, 26 and 27 Matrix: Climate impacts on activities, 28-1 Gains from damming rivers, 29-1 Rivers in nature, 29-2 Rivers in cities, 29-3 New design philosophy, 30-1 Shift of habitable areas, 30-2 Preventive static features, 32-2 Balance of release and binding of CO_2, 33-1 Division of world emissions, 37-1 Arrow of time, 37-2 Cyclic process, 38-1 Tai-Chi symbol, 46-1 Substitution of hot areas, 46-2 Summer temperatures in the North, 46-3 Directions to go for cooling, 47-1 Megapattern: Towards the poles, 47-2 Megapattern. Towards coasts, 48-1 Global spread of humans, 48-2 Routes of the Vikings, 48-3 Discovery of sea routes, 49-1 Christaller Model, 49-2 The Silk Route, 50-1 Pattern: Centrifugal force, 50-2 Pattern: Centripetal force, 50-3 Madrid at a geographical centre, 51-1 Pattern: Coast a linear centre, 51-2 Pattern: Coastal ribbon of habitation, 51-3 Pattern: Linear centre in a ribbon, 51-4 Pattern: Point centres at coasts, 51-5 Pattern: Pull to the interior, 51-6 Pattern: Centre at the geographical middle, 52-1 Shipping lane along Siberia 52-2 Shift in oil exportation, 53-2 Sami migrate with a herd, 53-3 A dynamic world, 53-4 Settlement in a sunny spot, 54-1 People migrate as birds do, 55-1 Need for economic unions, 55-2 Areas of destination?, 55-3 Friends and foes, 57-1 Patterns of migration from Europe, 57-2 Pattern of migration towards the poles, 59-1 Today's linear centre, 59-3 The ribbon widens, 59-4 Semi-global world, 60-1 Isolated world centres, 60-2 Global communication, 60-3 Flat Earth terminology, 60-4 A sphere terminology, 60-5 Aviation terminology, 61-1 Linear centre of Earth, 61-2 Widening of the ribbon, 62-4 Future Arctic shipping lanes, 63-1 Belts of social spaces, 63-2 Spheres of social spaces, 64-1 On a globe: All points equal, 64-2 Most landmass in the northern hemisphere, 64-3 North Pole: Future centre, 65-1 All points similar, 65-2 Ancient point centres, 65-3 Ribbon develops, 65-4 Cyberspace starts, 65-5 Ribbon splits in two, 65-6 North gains importance, 66-3 New habitable areas in the North, 67-1 Coastal development, 67-2 Moving inland, 67-3 Ribbon gains a linear centre, 67-4 Shipping: Coastal towns develop, 67-5 Land transportation: Interior gains, 67-6 Centre at the geographical centre, 69-1 The Arctic: A continuum, 69-2 Antarctica: Not a continuum, 69-3 The ribbon will split, 71-1 First Arctic lane, 71-2

Second Arctic lane, 71-3 Third Arctic lane, 71-4 Fourth Arctic lane, 72-2 Hudson Bay lanes, 72-3 Future Arctic lanes, 73-1 Lanes at Bering Strait, 73-2 China: No north access to the Pacific, 74-1 Symbolic "World Capital", 74-2 The gates into the Atlantic, 76-1 Sequence in changes, 76-2 Flat Earth: The centre most important, 76-3 Ribbon: It's linear centre most important, 76-4 Globe: Has no centre, 77-1 Today's world: No common focus, 77-2 The future: North pole a focus, 77-3 A mixture of ribbon and globe, 77-4 The semi-sphere worldview, 78-1 Horizontal and vertical relationships, 78-2 Growing importance of the sphere, 78-3 Semi-global world will take over, 79-1 Now: Lack of integration, 79-2 Future: Increased integration, 80-3 Barriers to global shipping, 81-1 Barriers to global railways, 81-3 Global high-speed train, 82-2 Future global shipping system, 82-3 Future: New global shipping circles, 83-1 Global high-speed train system, 83-2 Railways and their sea links, 83-3 Chicago: A main centre, 83-4 Beijing: A main centre, 83-5 Berlin: A main centre, 86-2 Future: Areas unfit for habitation, 87-2 Future: Areas good for habitation, 88-3 Ribbons around continental railways, 89-1 Arctic routes: Areas that gain or lose, 91-1 Europe: First industrial cities at coasts, 91-2 Land transportation: Inland linear centre, 91-3 Coastal town development, 91-4 Interior development, 91-5 Centre develops, 92-1 USA: Settled from coasts, 92-2 Interior core at the Great Lakes, 92-3 Migration to coasts and south, 93-1 Migration to coasts, 93-2 Migration to coasts, and north, 93-3 Midwest centre moves north, 95-1 Siberia: From west to east, 95-2 Siberia: Development inland, 95-3 Siberia: Ribbon develops at coast, 95-4 A new northern railway line, 95-5 Less ice: River corridors, 95-6 Centres at river-railway intersections, 96-2 Future structure of Central Asia, 96-3 World map composed of 60 triangular polyhedrons, 97-1 Arctic sea routes, 97-2 Megapattern: Toward the poles, 97-3 Megapattern: Toward coasts, 97-4 Megapattern: Toward central areas , 98-2 Europe's future frontiers, 99-1 EU: Brussels was close to centre, 99-2 EU: Centre of Earth's landmass, 99-3 Europe: Berlin close to its centre, 99-4 The power of the centre, 100-1 The Arctic: All time zones become one, 100-2 Vertical relations grow in importance, 100-4 Arctic rivers: New access to the continents, 101-1 Areas within reach of high-speed trains, 101-2 Section through a high-speed tube, 106-2 Fifteen regions of the world, 110-1 Today's world spatial system, 110-2 The future world spatial system, 110-3 Areas with shortage of water in 2025, 111-1 Areas that will lose because of Arctic shipping, 112-2 Arctic oil areas, 112-3 World oil areas, 113-1 Fifteen regions of the world, 119-1 New oil partnerships, 119-3 Short distances from Barents Sea oil fields, 120-1 Shallow Arctic continental shelf, 120-3 North of conflict in WW II, 121-1 Arctic shipping north of conflict, 121-2 Conflict areas of the future, 122-3 The Arctic: An area of common interests, 123-1 Barriers to migration, 125-1 Ice retreats, Arctic routes open, 128-4 CO_2 emission scenarios, 128-5 Arctic precipitation increase, 129-1 Extension of trends to 2150, 131-1 Passages north and south of the continents.

Two of the *best sources for illustrations* and *points made* in the book were: *ACIA, Impacts of a Warming Arctic,* and *Arctic Climate Impact Assessment.* Cambridge University Press, 2004 and 2005.

The following illustrations were sourced from these books: 14-1, 14-2, 14-3, 18-2, 19-2, 20-2, 20-3, 125-2, 126-1, 126-2, 128-1, 128-2, 128-2, 130-2, 134-3, 134-4, 139-3.

Another main source for illustrations was *Vital Arctic Graphics: People and Global Heritage on Our Last Wild Shores.* Joint publication by UNEP, GRID-ARENDAL, WWF and CAFF.

The following were taken from the indicated sources:

70-3: *Protected areas of the Arctic.* Source: World Protected Areas Database, UNEP-WCMC (2005).

134-3: *Arctic indigenous peoples.* Adapted from map by W.K. Dallmann published by the Stefansson Arctic Institute in the 2004 Arctic Human Development Report. Data and information compiled by W.K. Dallmann, Norwegian Polar Institute and P. Schweitzer, University of

Alaska Fairbanks. Further modified after expert feedback.

134-4: *States, organizations and strategic issues in the Arctic: People across borders.* Sources: Arctic Council; Norwegian Polar Institute, Permanent Participants of the Arctic Council, map compiled by Winfried Dallmann; Global Security, Washington DC; Department of Foreign Affairs and International Trade of Canada; International Boundaries Research Unit Database, University of Durham. Updated with expert feedback.

137-1: *Topography and the bathymetry of the Arctic.* Source: Digital Chart of the World; Institut Géographique National Français (IGN). AMAP, 1997. Arctic Pollution Issues: A State of the Arctic Environment Report.

138-1: *Vegetation distribution of the Arctic and the Sub-Arctic.* Sources: S. Tuhkanen, A Circumpolar System of Climatic Phytogeographical Regions, 1984; Acta Botanica Fennia 127, 50pp.; L. Hämet-Ahti, Subdivision of the Boreal Zone, Fennia 159:1 69-75, 1981; Atlas Arktiki USSR 1985 – 114-121; Atlas Mira: USSR 1964; 6667DCW 1996-8 (digital charts of the world); CAFF Habitat Conservation Report no. 5. This map was first published in CAFF 2001, Arctic Flora and Fauna, Status and Conservation. Adapted from map by Seppo Kaitala and Juhana Nieminen (University of Helsinki).

139-1: *Cumulative impacts of development in the Arctic.* Source: Modified from analysis published in the GEO3 Global Environment Outlook (2002).

139-2: *Scenarios of development.* Source: Modified from analysis published in the GEO3 Global Environment Outlook (2002).

141-1: *Industrial development in the Arctic.* Sources: United States Geological Survey (USGS); AMAP 1997, 1998 and 2002; CAFF, 2001; UNEP/World Conservation Monitoring Centre (WCMC); United States Energy Information Administration (EIA); International Energy Agency (IEA); Barents Euro-Arctic Council (BEAD); Comité professionel du pétrole (CPDP), Paris; Institut français du pétrole (IFP), Paris; National Oceanic and Atmospheric Administration (NOAA); The World Bank; Alaska Department of Environmental Conservation, Division of Spill Prevention and Response; United States Coast Guard (USCG).

146-2: *Population distribution and indigenous peoples in the Arctic. Sources.* United States: US Census Bureau, 2002, and United States Department of Commerce 1993; Canada: Statistics, 1995 and 2002; Greenland: Statistics Greenland, 1994 and 2002: Faroe Islands: Faroe Islands Statistics, 2002; Iceland: Statistics Iceland, 2002; Norway: Statistics Norway, 2002; Sweden: Statistics Sweden, 2002: Finland: Statistics Finland, 2002; Russia: State Committee for Statistics, 2003; Republican Information and Publication Centre, 1992: State Committee of the Russian Federation for Statistics 1992. AMAP, 1998. AMAP Assessment Report: Arctic Pollution Issues. AMAP, 1997. Arctic Pollution Issues. A State of the Arctic Environment Report. Stefansson Arctic Institute, 2004. Arctic Human Develop-ment Report.

146-3: *The largest economies in the Arctic.* Source: Stefansson Arctic Institute, 2004. Arctic Human Development Report.

147-1: *Indigenous settlements in the Arctic.* Sources: United States: US Census Bureau, 2002 and United States Department of Commerce 1993; Canada, Statistics Canada, 1995 and 2002; Greenland: Statistics Greenland, 1994 and 2002; Faroe Islands: Faroe Islands Statistics, 2002; Iceland: Statistics Iceland, 2002; Norway: Statistics Norway, 2002; Sweden: Statistics Sweden, 2002; Finland, Statistics Finland, 2002; Russia, State Committee for Statistics, 2003; Republican Information and Publication Centre, 1992; State Committee of the Russian Federation for Statistics 1992 ; World Wildlife Fund (WWF) Norway.

On the origin of the photos used in this book:
For a small publisher, *the Internet* offers possibilities to use illustration material that otherwise would be too costly. This advantage applies both for licensing fees and the costs of searching in traditional picture archives. On the Internet today there are *image banks* that offer *illustrations for use in books free of charge.* This book has benefited great-

ly from this new opportunity offered by the Internet, for which *we offer our best thanks*.

The invention of *digital photography technology* and the fact that such cameras have become widely owned by the general public is another reason that has made possible this extensive development of image banks on the Internet.

For the purposes of this book, images were only needed to express general impressions of the themes that the book deals with. These themes included, for instance, barren areas of the North, spreading deserts and improving environmental conditions. Also pictures that are general in nature were needed to give a visual sense of past periods, technological advances and future visions. The small pictures provided by the image banks were sufficient in size and quality for the general purposes of the visualizations needed for this book.

The image banks mostly used for this book were the *Wikipedia Image Bank,* the part of the *Flickr Image Bank* that has a Creative Commons licence, and the *Stock.xchng Image Database.* These image banks are created by the generosity of hundreds of thousands of people that have made pictures that they have taken available for use for everybody. In addition *websites of many official* and *governmental institutions* offer image material in the public domain, including NASA *(National Aeronautics and Space Administration)* and the USGS *(United States Geological Survey).*

The Stefansson Arctic Institute in Akureyri, Iceland (Photographer: Níels Einarsson) offered us with many pictures and other valuable assistance. The pictures are on the following pages: 20-1 Lifestyle changes in the Arctic, 41-1 Arctic societies become modern, 70-1 Iceberg, 85-1 Cold areas will become hospitable, 86-1 Ice retreats from coasts, 136-1 An Arctic husky, 137-2 Fishing increases with less ice, 147-3 Eskimo settlements.

The National Museum of Iceland. The following picture: 103-1 Ice at a town. From the book *Ljósmyndir Sigfúsar Eymundssonar* p. 203 (Photographer S. Eymundsson).

National Archives and Records Administration, Maryland, USA. The following picture: 120-3 Warships in Hvalfjord. From the book *Ísland í hershöndum* (2002), p. 217.

The Norwegian Polar Institute supplied the following picture: 124-3 Scheme to assess environmental impacts. From the book *Northern Sea Route – Future & Perspective,* p. 207 (1996).

The project II.4 *Mapping of Valued Ecosystem Components – III: Marine Mammals* by S. E. Belikov et al. (1995), supplied two pictures: 145-1 Distribution of breeding King Eider, p. 218, 145-2 Distribution of walrus and walrus haulouts, p. 219. Taken from the book *Northern Sea Route – Future & Perspective* (1996).

A source for one picture was: U.S. Congress. Senate. Committee on Armed Services. Department of Defense. *Authorization for Appropriations for Fiscal Year 1980, Part 6:* 74-3 Iceland's strategic position. Published in the book *GIUK-hlidid* by Gunnar Gunnarsson, Öryggismálanefnd, rit 1, p. 52 (1981).

The *Copenhagen Consensus* project supplied one chart: 31-1 Final overall ranking. Published in *Global Crises, Global Solutions,* p. 606 (2004).

Images by Jacque Fresco: 115-1 Undersea world, p. 154, 115-3 Cybernated Complex, cover, 116-1 Global communication, p. 134. Published in the book *The Best That Money Can't Buy* (2006).

One picture was taken from the book *Ancient People of the Arctic* (2001): 135-1 Climatic change in Arctic Canada since the last Ice Age, p. 108.

A permit was gained to take six pictures from the book *North Meets North* (2006): 62-1 Comparing shipping routes, p. 22, 62-2 Sixth generation of container ships, p. 27, 62-3 Comparing shipping distances, p. 25, 102-1 Samskip crane, p. 36, 102-2 Sundaharbour, Reykjavík, p. 38 (Photographer Haukur Snorrason), 107- Oil tanker Tempera, p. 30.

From the book *Saga Reykjavíkur 1870-1940:* 68-3 Ship in Reykjavík Harbour, p. 327 (Photographer Magnús Ólafsson).

Index of Terms and Acronyms

Acronyms

AC *Arctic Council*

ACAP *Arctic Council Action Plan*

ACIA *Arctic Climate Impact Assessment, a study conducted under the auspices of the Arctic Council*

AHDR *Arctic Human Development Report*

AMAP *Arctic Monitoring and Assessment Programme*

AMTW *Arctic Marine Transport Workshop*

AOSB *Arctic Ocean Science Board*

ARCDEV *Arctic Demonstrations and Exploratory
Voyage*

ASOF *Arctic and Sub-Arctic Ocean Flux Programme*

CAFF *Conservation of Arctic Flora and Fauna*

EEA *European Environment Agency*

EPB *European Polar Board*

EPPR *Emergency Prevention, Preparedness and
Response*

EU *European Union*

FARO *Forum of Arctic Operators*

GHGS *Greenhouse gases*

IASC *International Arctic Science Committee*

ICARP II *Second International Conference for Arctic
Research Planning*

IIED *International Institute for Environment and
Development*

INSROP *International Northern Sea Route
Programme*

IPCC *Intergovernmental Panel of Climate Change*

IPY *International Polar Year 2007 - '08*

IRIS *Ice Riding Information for Decision Making in
Shipping Operations*

NCDC *National Climatic Data Center (USA)*

NSR *Northern Sea Route*

OMAE *Offshore Mechanics and Arctic Engineering*

PAME *Protection of the Arctic Marine Environment*

SEA *Strategic Environmental Assessment*

UN *United Nations*

UNDP *United Nations Development Programme*

UNEP *United Nations Environment Programme*

UNFCCC *United Nations Framework Convention on
Climate Change*

USEPA *U.S. Environmental Protection Agency*

USGS *Unites States Geological Survey*

WEF *World Economic Forum*

WMO *World Meteorological Organization*

WRI *World Resources Institute*

WWF *World Wide Fund for Nature*

Index of Persons and Places

Summary   *"How the World will Change – with Global Warming"*

Countless reports and books have already been published on the climatic aspects of global warming. What these climatic changes mean in terms of the natural environment has also been extensively described by scientists. What these changes will mean in terms of human settlement and settlement structures has, until this book, been little described.

The book starts by describing how, for over one hundred years, the climate on Earth has been warming. This has already led to enormous changes in the bio-systems and weather patterns of the globe. The book proceeds by describing manifold projects, like Kyoto, that are underway to reduce the greenhouse gas emissions that cause the warming. Many climatologists predict, however, that in spite of efforts to check the warming it can not be stopped. Scientists, however, fully realize that it is of highest importance to try to reduce the speed of the warming in order to reduce its catastrophic consequences, mostly in the temperate and tropical regions of the world.

The warming and cooling of Earth's climate is not new. Historically humankind has responded to these fluctuations in various ways, largely in times of warming by spreading settlements and activities towards the poles, to higher elevations or to coasts just as other life systems do. Under conditions of excessive warming only migration towards the poles will suffice. Today's world of fixed settlements and fixed political borders will, however, make this type of migration more difficult than it has been in the past.

It is almost embarrassing for the people of the North to admit that the warming of the globe brings several benefits to the now cold areas of the globe. The climate is already much more comfortable there and agricultural yields are greatly improved in areas where there is enough rain. These gains – that are hardly ever heard of in public discussion – prompted the author to study, in a balanced way, what the losses and gains from the climatic changes will be for the globe as a whole. The book presents maps that show the areas that will be hardest hit – and also, areas where conditions are improving, and will improve still further. These cover huge areas in N Canada, Alaska, Scandinavia and Siberia where climatic conditions will improve greatly and where enormous oil and other resources have already been found and will be increasingly accessible.

Of most consequence for the world is the prediction that as soon as within thirty to forty years the Arctic Ocean ice will have retreated so much – particularly during the summer – that major shipping lanes between the North Atlantic and the North Pacific areas will have developed. These shipping routes will be much shorter than the present shipping routes between continents and world oceans. In addition the ever increasing size of ships means that more and more ships cannot pass through the narrow Suez and Panama canals and therefore have to go the very long routes beyond the southern tips of Africa and South America.

If the warming of the globe becomes extensive many areas in the southern and central areas of the globe will become uninhabitable. Such an excessive warming of the globe will, however, at the same time mean that the Arctic will become so much warmer that it will have a climate similar to that of Scandinavia today. A possible solution to the problems of the areas that are stricken by the impacts of global warming is the migration to the, by then, benign areas of the Arctic and Sub-Arctic.

The author recognizes the fact that the world's climate has changed many times in the past and that the present warming trend is not new. However, he underlines the importance of the international agreements for reducing greenhouse gas emissions, which are largely to blame for the current warming trend. He also promotes the view that, even in the face of the very many negative consequences, we need to maintain a positive attitude towards the changes that are coming upon us. Foreknowledge can help us all to plan for and adjust to, not only the negative aspects of the changes, but also the beneficial ones.

Epilogue

The work on this book can be said to have started in the 1980´s as the author was working on his PhD degree in environmental planning at University of California, Berkeley. This environmental type of approach to planning was then in its formative stages at Berkeley. The university is known for its extensive work within the environmental field. UC Berkeley was therefore a very good university to gain knowledge on how human habitat and the natural environment interrelate. My most important professors in this field were Joe McBride, Luna Leopold and Michael Laurie. Unfortunately, the last two have passed away. I owe all of them a great deal for their help and guidance.

After the author returned to his home in Reykjavík, Iceland, he was soon given a part-time position at the University of Iceland in the Faculty of Engineering, as there was and is no department of planning in any Icelandic university. The environmental and civil engineers, as part of their studies, need an introduction to environmental and planning concerns. The author's position soon developed into a full-time position and a tenured professorship. From working with engineers the author has gained improved insight into engineering subjects, including transportation, soil stability, erosion, environmental impact assessment and natural hazards. Most benefits for the present work have been gained in collaboration with Professors Ragnar Sigbjörnsson, a natural hazards expert, and Birgir Jónsson, a geo-engineer. My best thanks to them.

As the author was invited as a Farrand Visiting Professor to teach at his old LAEP Department at Berkeley in the autumn of 2004, the course was dedicated to the impacts of global warming in the Bay Area, the first course on the subject at the department. That year was the second of the two years during which Iceland held the chairmanship of the Arctic Council. At the ministerial conference in Reykjavík in November 2004 the main findings of the work of 300 scientists on "Arctic Climate Impact Assessment" (ACIA) were announced, catching the attention of the world media. The chairman on behalf of Iceland in the group of senior officials for this work was Ambassador Gunnar Pálsson and the Icelander Snorri Baldursson, Deputy Director of the Icelandic Natural Sciences Institute, was a member of the Assessment Integration Team for the ACIA report. Many Icelandic specialists took part in the research, Iceland's senior expert on sea ice, Thór Jakobsson, being one of them. They have all been very helpful during the work on this book. My talks also go to Eiríkur Benjamínsson MD and Albert Jónsson, now Iceland's Ambassador to the USA, my co-author of a book on a changing world in 1995.

Rannís – The Icelandic Centre for Research organized four seminars in 2005 on Arctic matters that were of great help. The author is a member of the web journal EJSD, published by Nordregio in Stockholm. In two editorial meetings, in Milan and Paris, the author presented his thoughts on the subject and was given valuable feedback.

The work on the layout and the final text of the book started in January 2006. Björk Ingadóttir, Sigrídur Thorarensen, Thorgerdur Sveinbjarnardóttir and Gunnhildur Jóhannsdóttir gave valuable assistance on the text. The editing and language correction was in the hands of Dr Terry G. Lacy, whose work on the text was invaluable. Assistants on the visual aspects of the book were Karl Jóhann Magnússon, Salvar Th. Sigurdsson, Benedikt Ingi Tómasson and Kjartan Jónsson. My best thanks to them all.

The financial support for the production of the book was provided by the Baugur Cultural Fund. I am also very indebted to a great number of specialists who have given advice and read sections of the book. The author, however, alone carries the ultimate responsibility. My thanks also go to the scholars who were so kind as to read the book manuscript and write the comments that are printed on page 6.

Trausti Valsson

General References and Websites

ACIA *Arctic Climate Impact Assessment.* Cambridge University Press, 2005

Acta Borealia. *Nordic Journal of Circumpolar Societies. Volume 22.* UK, 2005

AHDR *Arctic Human Development Report.* Stefansson Arctic Institute, Akureyri, Iceland, 2004

Aidt, Leif: *Graenland.* Reykjavík, 1996

Barnett, Thomas P. M: *The Pentagon's New Map – War and Peace in the Twenty-First Century.* NY, 2004

Bernes, Claes: *Warmer World – The Greenhouse Effect and Climate Change.* Sweden, 2003

Brown, Lester R. et al: *State of the World 1994. A Worldwatch Institute Report on Progress toward a Sustainable Society.* USA, 1994

Burton, Rosemary: *Travel Geography.* London, 1991

Buzan, Barry and Little, Richard: *International Systems in World History – Remaking the Study of International Relations.* NY, 2000

Carr, Michael: *New Patterns: Process and Change in Human Geography.* UK, 1997

Center for Oceans Law and Policy: *International Energy Policy, the Arctic and the Law of the Sea.* The Netherlands, 2005

Crichton, Michael: *State of Fear.* USA, 2004

Diamond, Jared: *Collapse – How Societies choose to Fall or Succeed.* USA, 2005

Doxiadis, C.A. and Papaioannou: *Ecumenopolis – the Inevitable City of the Future.* Greece, 1974

The Economist: *The World in 1994.* London, 1993

The Economist: *The World in 1996.* London, 1995

The Economist: *The World in 2006.* London, 2005

Encyclopedia Britannica: *Britannica Atlas.* USA, 1977

European Environment Agency: *Environment in the European Union at the Turn of the Century. Offprint: Soil degradation.* Copenhagen, 1999

Fleischhauer, Mark: *Klimawandel, Naturgefahren und Raumplanung.* Dortmund, Germany, 2004

Fuller, R. Buckminster: *Inventory of World Resource: Human trends and needs.* Illinois, 1971

George Marshall Institute: *Climate Issues & Questions.* Washington, DC 2004

Global Biodiversity Assessment – Summary for Policy-Makers. Published for UNEP by Cambridge University Press, 1995

Gore, Al: *An Inconvenient Truth. The Planetary Emergency of Global Warming and What We Can Do About It.* NY, 2006

Guinnes, Paul: *Globalization.* London, 2003

Gunnarsson, Gunnar: *GIUK-hlidid.* Reykjavík, Iceland 1981

Hagbarth Publications: *Structural Change in Europe – New Northern Knowledge.* Germany, 2002

Hall, Peter: *Cities in Civilization.* London, 1999

Harrison, Paul and Pearce, Fred: *AAAS Atlas of Population & Environment.* London, 2000

Hassol, Susan Joy: *Impacts of a Warming Arctic – Arctic Climate Impact Assessment.* Canada, 2004

Hedegaard, Lars and Lindström, Bjarne (editors): *The Nebi Yearbook 1998 – North European and Baltic Sea Integration.* NY, 1998

Held, David (editor): *A Globalizing World? – Culture, Economics, Politics.* London and NY, 2004

Huntington, Samuel P: *The Clash of Civilizations and the Remaking of World Order.* NY, 1996

Icelandic Ministry for the Environment: *Ísland og Loftslagsbreytingar af Mannavöldum. Skýrsla umhverfisrádherra.* Reykjavík, 1997

Icelandic Ministry of Foreign Affairs: *Fyrir stafni haf – Taekifæri tengd siglingum á nordurslódum.* Reykjavík, 2005

Icelandic Ministry for Foreign Affairs, Iceland: *North Meets North – Navigation and the Future of the Arctic,* Reykjavík, 2006

Icelandic Ministry for the Environment: *Vedurfarsbreytingar og afleidingar theirra – Skýrsla vísindanefndar um loftlagsbreytingar.* Reykjavík, 2000

Ilbery, Brian W.: *Western Europe – A Systematic Human Geography.* NY, 1990

Institute of the North: *Arctic Marine Transport Workshop.* UK, 2004

Intergovernmental Panel on Climate Change: *The Regional Impacts of Climate Change – an Assessment of Vulnerability*. UK, 1998

International Energy Agency: *World Without Outlook*. France, 1995

Joint Seminar and Workshop of the Barents Regional Council and the Northern Forum: *Envisioning the Northern Dimension*. Rovaniemi, 1999

Jónsson, Albert: *Ísland Atlantshafsbandalagid og Keflavíkurstödin*. Reykjavík, 1990

Jónsson, Albert: *Vígbúnadur og Fridunarvidleitni við Indlandshaf*. Reykjavík, 1982

Jónsson, Björn: *Vilhjálmur Stefánsson landkönnudur*. Reykjavík, 1995

Kennedy, Paul: *Preparing for the Twenty-First Century*. NY, 1993

Kidron, Michael and Segal, Ronald: *The New State of the World: A Pluto Press Project*. NY, 1984

Kitagawa, Hiromitsu (editor): *Northern Sea Route – Future & Perspective. Proceedings of the INSROP Symposium Tokyo '95*. Ship & Ocean Foundation. Tokyo, 1996

Kotkin, Joel: *The New Geography: How the Digital Revolution is Reshaping the American Landscape*. New York, 2000

Laszlo, Ervin et al: *Goals for Mankind*. New York, 1977

Lean, Geoffrey, Hinrichsen, Don and Markham, Adam: *Atlas of the Environment*. London, 1990

Legro, Jeffrey W: *Rethinking the world – Great Power Strategies and International Order*. NY, 2005

Leonard, Mark: *Why Europe will Run the 21st Century*. UK, 2005

LHS GEMS: *Global Warming and the Greenhouse Effect*. USA, 1990

Lomborg, Bjørn (editor): *Global Crises, Global Solutions*. Cambridge, 2004

Lomborg, Bjørn: *Hid sanna ástand Heimsins*. Akureyri, 2000

Lomborg, Bjørn: *The Skeptical Environmentalist – Measuring the Real State of the World*. Cambridge, 2005

Makridakis, Spyros G: *Forecasting, Planning and Strategy for the 21st Century*. London, 1990

Mariussen, Åge: *Nordic Institutions and Regional Development in a Globalized World*. Stockholm, 1998

Mumford, Lewis: *The City in History*. London, 1974

Murray, Warwich E: *Geographies of Globalization*. London and NY, 2006

Myers, Norman (editor): *Gaia – An Atlas of Planet Management*. London, 1984, 1993

Myers, Norman and Kent, Jennifer: *The new Gaia Atlas of Planet Management*. London, 2005

Naisbitt, John and Aburdene, Patricia: Megatrends 2000: *The next 10 years. Major Changes in your Life and World*. London, 1990

Naisbitt, John: *Megatrends. Ten new Directions Transforming our Lives*. NY, 1982

Naisbitt, John: *Global Paradox – The Bigger the World Economy, the More Powerful Its Smallest Players*. NY, 1994

National Assessment Synthesis Team: *Climate Change Impacts on the United States : The Potential Consequences of Climate Variability and Change – Overview*. UK, 2000

National Centre for Educational Materials: *Landabréfabók*. Reykjavík, 1991

National Geographic: *Almanac of Geography*. Washington, D.C., 2005

Naturopa, theme issue: *Nature: The Common Heritage of Humankind. No. 91*. Brussels, 1999

NORD: *Arctic Challenges*. Stockholm, 1993

Norden: *Nordisk regionalpolitisk samarbeidsprogram*. Copenhagen, 2005

Norræna rádherranefndin: *Náttúra Nordurlanda – Forsendur og meginreglur náttúruverndar*. Nordic Council of Ministers, Copenhagen, 1995

North – The Journal of Nordregio, theme issue: *Future Challenges to Nordic Regional Policy*. Volume 10. Denmark, 1999

Nuttall, Mark (editor): *Encyclopedia of the Arctic I-III*, 2005

Novgorod, Veliky: *Northern Veche – Proceedings of the Second NRF Open meeting*. Russia, 2002

Overy, Richard: *Atlas of 20th Century*. London, 2005

Pittock, A. Barrie: *Climate Change – Turning up the Heat*. London, 2005

Proceedings: *International Navies. Volume 132/3/1, 237*. USA, 2006

Psomopoulos, Panayis (editor): *Ekistics: the Problems and Science of Human Settlement – Urban Networking in Europe – I. Volume 58*. Greece, 1991

Psomopoulos, Panayis (editor): *Ekistics: the Problems and Science of human Settlements – Urban Networking in Europe – II. Volume 59*. Greece, 1992

Sale, Richard: *To the Ends of the Earth: The History of Polar Exploration*. London, 2002

Schram, Gunnar G: *Ísland á nýrri öld. Hugleidingar 22 Íslendinga vid upphaf nýs árthúsunds*. Reykjavík, 2000

Scientific American, Special Issue: *Our Ever Changing Earth. Volume 15, Number 2*. NY, 2005

Scoffham, Stephen: *World Environment Atlas*. Scotland, 1991

Sigfússon, Thór: *Örríki á Umbrotatímum. Leidabók fyrir nýja öld*. Reykjavík, 1997

The Nordic Council: *Northern Europe's Seas: Northern Europe's Environment*. Stockholm, 1989

The World Bank: *World Development Report 1991: The Challenge of Development*. USA, 1991

Toffler, Alvin: Future Shock. London, 16th printing. London 1979

United Nations: *World Statistics in Brief*. New York, 1979

Valsson, Trausti and Jónsson, Albert: *Vid aldahvörf. Stada Íslands í breyttum heimi*. Reykjavík, 1995

Valsson, Trausti and Jónsson, Birgir: *Ísland hid nýja*. Reykjavík, 1997

Valsson, Trausti: *Hugmynd ad fyrsta heildarskipu-lagi Íslands*. Reykjavík, 1987

Valsson, Trausti: *Planning in Iceland – From the Settlement to Present Times*. Reykjavík, 2003

Valsson, Trausti: *Geschichte und Zukunft Islands aus planerischer Sicht*. Island, Zeitschrift der DIG, Köln, 2006

Viotti, Paul R. and Kauppi, Mark V.: *International Relations and World Politics – Security, Economy, Identity*. New Jersey, 2002

Whitbeck, Ray H. And Thomas, Olive J: *The Geographic Factor – Its Role in Life and Civilization*. London and New York, 1970

Wolff, Michael: *Where We Stand – Can America make it in the Global Race for Wealth, Health and Happiness?* USA, 1992

World Tourism Organization: *National and Regional Tourism Planning – Methodologies and Case Studies*. London and New York, 1994

Websites

ACIA Scientific Report http://www.acia.uaf.edu

Asia-Pacific Partnership on Clean Development and Climate http://www.sustainibility.act.au/green-house

Arctic Climate Impact Assessment (ACIA) www.acia.uaf.edu and www.amap.no/acia

Arctic Demonstrations and Exploratory Voyage (ARCDEV) http://arcdev.neste.com

Arctic Marine Transport Workshop (AMTW) http://www.institutenorth.org and http://www.iascp.org

Arctic Operational Platform (ARCOP) http://www.arcop.fi

Centre for Science and Environment (India-based organisation) www.cseindia.org

Climate Options for the Long term (Dutch project) www.wau.nl/cool

CO2e.com www.CO2e.com

European Environment Agency (EEA) www.eea.eu.int

Friends of the Earth www.foe.co.uk

Fusion Program ITER www.iter.org

Global Environment Outlook www.unep.org/GEO/geo3

Greenpeace www.greenpeace.org

Hydrogen

www.fe.doe.gov/programs/fuels
Ice Riding Information for Decision Making in Shipping Operations (IRIS)
http://www.hut.fi/Units/Ship/Research/Iris/Public
ICE Routes
http://www.cordis.lu/transport/src/icer-outerep.htm
India Ministry of Environment and Forests
www.envfor.nic.in
International Institute for Environment and Development (IIED) www.iied.org
International Northern Sea Route Programme (INSROP) http://www.fni.no/insrop
Intergovernmental Panel on Climate Change (IPCC) www.ipcc.ch
International Polar Year 2007 - '08 (IPY) www.ipy.org
Marine Research Institute of Iceland www.hafro.is/symposium
NASA Earth Observatory http://earthobservatory.nasa.gov
National Climatic Data Center (NCDC) http://www.ncdc.noaa.gov/oa/ncdc.html
National Renewable Energy Laboratory www.nrel.gov
Norden www.norden.org
Nordic Centre for Spatial Development www.nordregio.se
Northern Sea Route (NSR) http://www.beac.st
Norwegian Ministry of Foreign Affairs http://www.odin.dep.no/
Offshore Mechanics and Arctic Engineering (OMAE) http://ooae.org/index.htm
Protection of the Arctic Marine Environment (PAME) http://www.pame.is
RealClimate www.realclimate.org
Royal Institute of International Affairs (UK) www.riia.org
Shell Hydrogen www.shell.com/hydrogen
Stefansson Arctic Institute (SAI) http://www.svs.is
The Climate Group www.theclimategroup.org
U.S. Environmental Protection Agency (USEPA)

http://yosemite.epa.gov/oar/globalwarming.nsf
UNEP/GRID-Arendal www.grida.no
United Nations Centre for Human Settlements (UNCHS) www.unchs.org
United Nations Development Programme (UNDP) www.undp.org/cc
United Nations Environment Programme (UNEP) www.unep.net
United Nations Framework Convention on Climate Change (UNFCCC) http://unfccc.int/
Unites States Geological Survey (USGS) http://search.usgs.gov/
Vital Arctic Graphics.pdf www.grida.no
World Bank Data and Statistics www.world-bank.org/data/maps/maps.htm
World Climate Report http://www.nhes.com/home
World Economic Forum (WEF) http://www.weforum.org/site/homepublic.nsf
World Energy Agency www.worldenergy.org
World Meteorological Organization (WMO) http://www.wmo.ch/web-en/index.html
World Resources Institute (WRI) www.wri.org/
World View of Global Warming www.worldviewofglobalwarming.org
World Wide Fund for Nature (WWF) www.panda.org
Worldwatch Institute (US) www.worldwatch.org/

Glossary

1) ARCTIC AND GLOBAL ISSUES

ACIA Arctic Climate Impact Assessment, a study conducted under the auspices of the Arctic Council.

Agenda 21 The UN Rio Conference of 1992 established a global agenda for the 21st century on environmental matters.

Antarctica The large island-continent at the South Pole. (See figure on page 9).

Aquifer Water-bearing formation capable of yielding exploitable quantities of water..

Arctic Council An association of the eight countries that border the Arctic area: Russia, the USA (because of Alaska), Canada, Denmark (because of Greenland), Iceland, Norway, Sweden and Finland.

Asia-Pacific Partnership (A6) A partnership of the USA, China, Australia, India, South-Korea and Japan on Clean Environment and Climate.

Biodiversity The number of different species or functional groups. Biodiversity is considered to be a resource that needs to be preserved, as proclaimed in *Agenda 21*.

Brundtland Report Alternate name for the UN report *Our Common Future,* published in 1987 in preparation for the Rio Conference, named for its chairman G. Brundtland.

Climatic change Changes in some aspects of the global climate. See *global warming.*

Ecotax A tax used for influencing human behaviour to follow an ecologically benign path.

El Niño An irregular variation of an ocean current that flows off the west coast of South America, raising sea surface temperatures off the coast of Peru.

Geopolitics Politics that deal with global scale issues of a strategic or military importance.

Global balance The essential balance of CO_2 and O_2 in the atmosphere. See *greenhouse gases.*

Global economy The economy of the globe as a whole.

Global evolution Study of global, spatial evolution is given rigour by identifying historic trends and megapatterns.

Global structures See *structure* and *infrastructure* in 2).

Global warming A term used to signify an array of climatic changes happening in the world today. The warming of the global atmos-phere can be seen as a catalyst for these changes. Global warming and cooling have often happened before in the geological history of Earth. The present warming is believed to be caused by anthropogenic release of *greenhouse gases.*

Global world A world that functions as a globe in a *topological* sense. See 2).

Globalization The increasing interconnectedness of the world; economically, culturally and politically, fundamentally powered by advances in transportation and communication. See *cyberspace in* 2).

Greenhouse gases (ghgs) Gases that absorb infrared radiation in the atmosphere. In nature the release and absorption of ghgs are normally in balance, but the emissions released by anthropogenic activity can no longer be absorbed fully by nature, so ghgs are gradually accumulating in the atmosphere, contributing to global warming. See *global balance.*

Intergovernmental Panel of Climate Change (IPCC) UN organ that has the task of studying *global warming* and projecting how it will proceed and what influence it will have.

International Polar Year (IPY) Takes place from March 2007 to March 2008.

Johannesburg Conference UN environmental conference held in 2002, 10 years after the *Rio Conference,* to assess progress since then and to draft a future direction.

Permafrost Perennially frozen ground. Mostly located in the polar areas.

Rio Conference UN conference held in 1992 to discuss and sign *Agenda 21,* a global environmental program for the 21st century.

2) PLANNING–SPACE–PHILOSOPHY

Active time The chemist I. *Prigogine* demonstrated how the forward direction of time, arrow of time, plays an active role in many chemical processes. In most branches of science the importance of the active passing of time is being discovered.

Airborne city A city that has come to be because of travel by air. Often located deep within countries.

Catastrophe areas Areas that are likely to be threatened with natural disasters such as floods, avalanches, landslides and earthquakes. See *hazards* and *risk assessment.*

Complementarity If two binary aspects work together, like city and nature and male and female, they enhance each other so that an extra value is produced.

Central Place Theory Describes rules that concern central places in planning, e.g. on how a hierarchy of centres forms a system of hexagonal grids as in crystals. (See figure on page 49). Often named for its discoverer, *Christaller Model.*

Climatic change Changes in some aspects of the climate. See global warming.

Cyberspace The electronic space of today that has no physical dimensions.

Development planning On a global scale the term is mostly used for inducing positive developments in developing countries or regions.

Dynamic forces Forces that impact settlement patterns. They divide into: *forces that attract* (resources, transportation facilities…) and *forces that repel* (crime, pollution, excessive heat…). (See figure on page 51).

Dynamic pattern The way spatial developments happen in a dynamic way. See *pattern.* The term static pattern is used to signify a fixed geometric configuration such as a *settlement pattern.*

Dynamic world A world that is in flux and constantly responding to changes. See the opposite: A *static world.*

Dynamism, types of The four main types of dynamism that are at work in an area: *Centripetal force* (a pull towards the centre), *centrifugal force* (a pull towards the rim), *forces of exchange* (exchange of goods and manpower) and *complementary force* (exchange with a complementarity effect). (See figure on page 51).

Ecology The science of how organisms work and interrelate with each other and with the environment. The term *human ecology* is defined as a parallel.

Entropy The tendency of a closed system to move from a less to a more probable state.

Environmental Impact Assessment (EIA) A formalized procedure to assess environmental, social and economic impacts. Used for large-scale or dangerous constructions or plans. Also, recently, to assess the impact of *climatic change.*

Environmental management Managing an area or a region according to environmental needs and considerations. See *environmental planning.*

Environmental planning A planning method that has the goal of making natural conditions and environmental principles a foundation for planning. See *environmental management.*

Environmental principles There are four main legal environmental principles: *User Pays Principle,* that those who use an environment or a resource shall pay for protecting or maintaining its qualities. *Polluter Pays Principle,* that those who pollute shall pay for the damage and its rectification. *Co-operation Principle,* that disagreements shall be resolved among the parties involved. *Pre-cautionary Principle:* This principle orders that plans or projects shall not be allowed, or be continued, if environmental harm might result.

Exchanger A centre that facilitates exchanges. Can, for instance, be a port, a transportation centre or cities in general. (See figures on pages 51 and 96).

Forces, field of Occurrences in a region, or on the globe, that result from *dynamic forces* at play in an area.

Future studies Studies on what the future is likely to hold. Methods employed include expert opinion in various fields, statistical methods etc.

Hazards Types of dangers, e.g. created by natural occurrences, such as earthquakes, floods, tsunamis, heat spells and droughts. (See a matrix on page 42-43). See also *catastrophe areas* and *risk assessment.*

Infrastructure Internal structure in the world, a country, region or town. There are social, economic and technological infrastructures. The most important infrastructures are the transportation infrastructures. Global infrastructures cover the whole globe.

Integration Making a whole out of regions economically and socially and in terms of *infrastructures.* The need for integration is constantly growing.

Interface areas An area where two unlike areas meet. Often they are viewed as trouble areas, where they actually have a huge potential for sharing and creating complementary values.

Linear centre A *spatial system* where a line functions as a centre in a long stretched area, as often happens around transportation lines. A settlement area around a linear centre is called a *ribbon of habitation.*

Linear world The world today functions, in a *topological* sense, as a line or a ribbon around the globe. A *global world* is gradually taking over.

Megapattern A spatial pattern covering large geographic and time scales. Shows dynamic megatrends in a spatial development. (See figure on page 97).

Megatrend A general trend on a large scale, omitting small scale aberrations.

Node A connecting point in a structure or an *infrastructure.* Example: A port is a node in a shipping transportation structure. Such *exchangers* are of many kinds and have many ways of functioning.

Pattern The typical way a certain spatial development happens in an area. *Dynamic patterns* are activated by a pull or repulsion that originates in environmental, resource or activity features. Patterns are also shaped by landscape and other spatial features. See *forces* and *megapattern.*

Paradigm Ways of looking at things. Shared assumptions that govern an outlook of an epoch and approach to scientific endeavours. A more general term is *world view.*

Paradigm shift A shift from one *paradigm* to the next, where shared assumptions and approaches change.

Planning philosophy The main thoughts and attitudes that guide planning. See *environmental planning* for an example on a planning philosophy.

Ribbon of habitation On a global scale it is the habitable belt that goes around the globe. A more complete term would be a ribbon belt or a cylinder, because the ribbon around the world connects to form a belt. (See figure on page 61). This ribbon covers an area on both sides of the globe's linear centre that goes around Earth. Today the ribbon is about 10,000 km wide, which is half the distance between the poles. With *global warming* the ribbon of habitation will stretch into the polar areas.

Risk assessment Is carried out after a *hazard* in an area has been defined. Puts the risk at hand into the context of frequency of hazard, value in the area and what protective measures can be taken to reduce the risk.

Scenario A sequential description of how development is likely to take place. Often, as a policy has been formulated, a scenario is drafted in order to show how the resulting development is likely to unfold over time. See also *sequence development.*

Semi-global world The world of the future – with global warming – is going to be the semi-global world of the northern hemisphere. (See figure on page 65). See *global warming* and *linear world.*

Sequence development Towns often follow a sequence; a *pattern,* in development that starts with some primary function and proceeds in a sequence of steps. The sequence in the spatial development of countries or regions often starts at coasts because of shipping and later moves into the interior with developing inland transportation. (See figure on page 51).

Settlement pattern The typical way a certain dynamic, spatial development happens in an area. See *pattern.*

Settlement structure A form-system in a settlement. See *structure.*

Spatial systems Basic types of spatial systems that originate in the *topological* qualities of form: a one dimensional system (a linear system); a two dimensional system; and a three dimensional, or spherical, spatial system.

Static world The assumption that the world is, or should not, be changing. This results primarily from the poor awareness of the active nature of time, which is one of the fundamental characteristics of today's world view.

Strategic Environmental Assessment (SEA) The SEA is a further development of *Environmental Impact Assessment (EIA)* that takes this type of thinking to the higher level of strategies. The main idea is that environmental goals and concerns should be introduced at the initial stages of strategies, laws, plans and programs.

Structure A form-system in the world, in a country, in a region or a town. There are social, economic and technological structures. The most important structures are the transportation structures. *Global structures* cover the whole globe. See also *infrastructure.*

Sustainable development A principle that proclaims that the use of resources should not damage or reduce their capacity. Only in this way can their utilization become sustainable. This principle not only covers the natural environment but also the social and economic environments.

Topology A branch in mathematics that studies those properties of figures in space, which persist under all continuous deformations.

Trends Investigations are conducted into various types of trends, such as economic- and migration trends. Trend studies are used in *future studies.* See also *megatrend.*

World view Ways of looking at things. Shared assumptions that govern an outlook of an epoch and approach to scientific endeavours. A more specific term is *paradigm.*

Let's Embrace Change! The message of the book told as a poem (TV)

I A New Way of Life

Modern man
Is static in his ways.
This is in conflict
With the nature of the world.

Time changes everything,
But time is not accepted.
And we don´t accept the changes,
That time brings.

The world is changing,
Fast and profoundly.
Today, global warming
Is the catalyst of change.

Not to fight change
Is a lesson to be learned.
The new way of life;
Is to embrace change!

II Patterns of Change

From dust to dust;
The story of planets.
This same pattern
Applies to everything.

To a day-fly,
Not much changes.
The same holds true
For us humans.

Looking at history,
Continents change,
Organisms change –
Nothing is static.

Periods of warming,
Periods of cooling –
They come and go:
Patterns of change.

III To Fight Change

Man thinks:
I can control Nature!
And man wishes:
May nothing ever change!

On a small scale
Man is in control.
But in mega-events
He needs to reconsider.

Nations build levees
And gain land.
The fallacy becomes clear
With rising sea-levels.

Not to fight the oceans
Is the lesson here.
The oceans will rise –
It is ours to yield.

IV Change: A Disaster?

Change is trouble,
Change is hardship.
But we can rephrase:
Change is challenge!

We can gain from change,
Or let it defeat us.
The same with global warming:
We can gain from it.

There is a simple solution:
Yield to global warming,
Seek cooler areas,
Move toward the poles.

We have fouled our nest,
But new frontiers await;
Filled with opportunity,
For a new beginning.

V A New World

The Arctic from above
Is like a new planet.
Frost and isolation
Made it unknown.

A warm Arctic,
Is a new paradise;
A place to escape to,
From excessive heat.

Exploration and settlement,
Once a pride –
Have soured
From bad conduct.

Arctic settlement
Has to be planned
To make sure
It is sustained.

VI Objections

Some will say:
Let's stop global warming!
Sadly, not possible –
But it saves resources.

Others will say:
Let's create clean energy!
A good solution –
But not yet in sight.

As of now,
The warming will continue.
Let's draft a picture
Of northward migration.

Migration is a natural way
By warming and cooling.
Only static thinking
Makes it seem impossible.